47.50
7SE

OVERDUE FINES:

Walter F. Morofsky Memorial Library
W. K. Kellogg Biological Station
Michigan State University
Hickory Corners, Michigan 49060

JUN 1 9 1982

Man and Fisheries on an Amazon Frontier

Developments in Hydrobiology 4

Series Editor
H.J. DUMONT

DR W. JUNK PUBLISHERS   THE HAGUE–BOSTON–LONDON   1981

Man and Fisheries on an Amazon Frontier

MICHAEL GOULDING

Instituto Nacional de Pesquisas da Amazônia (INPA)
World Wildlife Fund

DR W. JUNK PUBLISHERS   THE HAGUE–BOSTON–LONDON   1981

Distributors

for the United States and Canada

Kluwer Boston, Inc.
190 Old Derby Street
Hingham, MA 02043
U.S.A.

For all other countries

Kluwer Academic Publishers Group
Distribution Center
P.O. Box 322
3300 AH Dordrecht
The Netherlands

Library of Congress Cataloging in Publication Data CIP

Goulding, Michael.
    Man and fisheries on an Amazon frontier.

    (Developments in hydrobiology ; 4)
    Bibliography: p. 121
    1. Fisheries--Brazil--Madeira River. 2. Fishes--
Brazil--Madeira River. 3. Madeira River (Brazil)
4. Fishery resources--Amazon Valley. I. Title.
II. Series.
SH236.G68   333.95'6'098112   81-8445
                                              AACR2

ISBN 90 6193 755 8 (this volume)
ISBN 90 6193 751 5 (series)

Cover design: Max Velthuijs

Copyright © 1981. Dr W. Junk Publishers, The Hague

All rights reserved. No part of this publication may be reproduced, stored in a retrieval system, or transmitted in any form or by any means, mechanical, photocopying, recording, or otherwise, without the prior written permission of the publishers Dr W. Junk Publishers, P.O. Box 13713, 2501 ES The Hague, The Netherlands.

PRINTED IN THE NETHERLANDS

Preface

The southwestern Amazon basin, centering on the Territory of Rondônia and the State of Acre, is symbolically if not exactly geographically, the Wild Wild West of Brazil's northern rainforest frontier. In Brazil the name Rondônia evokes exaggerated images of lawlessness, land feuding, and indigent peasants in search of a homestead. Despite the problems and the perception, the region has pushed ahead, in the view of the government, with large-scale deforestation and the establishment of cattle ranches and agricultural farms raising manioc, rice, bananas, and other cash crops. The mining industry has been launched with the exploitation of tinstone, and the recent gold rush has attracted thousands of miners that are sifting alluvial deposits along the rivers for the precious ore. In an energy-short world, the region boasts of its large hydroelectric potential waiting development in the rivers falling off the Brazilian Shield and draining into the Rio Madeira. Planners are optimistic that Rondônia's resources, once developed, will more than justify, at least in this corner of the rainforest frontier, the Economic Conquest of the Amazon. Sandwiched between the economic take-off and the dream, however, are the biological resources – the plants and animals – that must serve as sources of energy and food until human dominated ecosystems replace natural ones. These resources are, of necessity, being heavily attacked to support the shaky economy of the region, but they are very poorly understood in terms of potential productivity and proper management.

The most abundant of the natural resources that can be used as food in its wild state – and the one of main concern in this book – is fish. The rainforests themselves provide very little in the way of plants and animals that can be harvested on a large scale to feed many people. Most of the plant biomass is toxic and inedible, while rainforest vertebrates, the terrestrial and arboreal animals most acceptable as food in present day cultural standards, are too low in density to provide a significant protein source for large urban centers. Fishes offer several advantages, in terms of harvesting wild animals, that terrestrial and arboreal rainforest vertebrates do not. Per given area, at least during the low water season, they can be found in much greater concentrations. Some Amazon fish species are known to form schools of at least 100 tons, or 200,000 individuals, but it would be hard to imagine finding comparably sized rainforest bird flocks, monkey troups, or peccary bands. Rivers and floodplain lakes are also much easier to exploit for fishes than rainforests are for their vertebrates, as transportation is greatly facilitated by boat and canoe travel. These principal factors, and the cultural acceptance of fish as food, almost guarantee that the fish resource will be looked on for what it can furnish to the protein supply of the Amazon frontier.

In the case of the fisheries, the tentacles of the Rondônia frontier embrace most of the Rio Madeira, including its large headwater tributaries flowing out of Bolivia. This book represents an attempt to describe and quantify the nature of the Rio Madeira fisheries within the framework of the Rondônia frontier. I have been investigating fish ecology, fisheries, and human geography of the Rio Madeira region since 1976. My thrust has been to understand the Rio Madeira as a whole, and to elucidate the natural history of the region as part of the larger Amazon basin. It is contemplated that this work will serve as a trial run for a much larger treatise that will deal with all of the Amazon basin and its aquatic resources.

Before embarking on a description and analysis of the fisheries, I will first bring the study area into focus with physical, biological, and cultural portraits of the Rio Madeira valley. Chapter 3 discusses all of the types of commercial fisheries found in the Rio Madeira region, while Chapter 4 brings together the quantitative data in terms of fishing effort and yield. Chapter 5 offers an overview of the natural history of the Rio Madeira food fishes, and each of the species is discussed in some detail and accompanied by a photograph. The final chapter views the problems and prospects of the Rio Madeira fisheries.

With the exception of *standard length* and *fork length*, all terms used in the text should be self-explanatory if they are not defined when they appear. *Standard length* is the length of a fish taken from the tip of the snout to the base of the tail (caudal peduncle); this measurement is preferred for most fishes as it avoids uncertainty in determining lengths of fishes with often damaged caudal fin rays. *Fork length* is the length of a fish taken from the tip of the snout to the end of the median caudal rays, that is, to the end of the mid-section of the tail; this measurement is used herein for catfishes because many of the larger species have deeply forked tails and median caudal fin rays are seldom damaged, and thus lengths can be determined quickly and easily in the field.

MICHAEL GOULDING
Manaus, Amazonas, Brazil

Acknowledgements

I wish to express my especial thanks to Dr. Warwick Estévam Kerr for giving me the opportunity to be naturalist on the Rio Madeira, and to the Instituto Nacional de Pesquisas da Amazônia for employing me as research scientist and financing a large part of my work. To Dr. Thomas Lovejoy for his support of my projects, and to World Wildlife Fund for generous financing since 1979. Sr. Alfredo Nunes de Melo, *patrão* of the Rio Madeira fishermen, shared much of his natural history knowledge with me, and gave a helping hand in Rondônia. Sr. Vítor Aníbal de Lemos and the Superintendência de Desenvolvimento da Pesca (SUDEPE) gave logistical support in Rondônia that greatly aided the field work. From Manaus, Sr. Pedro Makiyama made sure that I received the necessary field equipment to keep the project rolling.

For criticism of an earlier, but somewhat different manuscript published in Portuguese, I thank Mr. Peter Bayley, Dr. Naércio Menezes, Dr. Nigel Smith, Dr. Thomas Zaret, Dr. Charles F. Bennett, and Dr. Rosemary Lowe-McConnell (who is also thanked for arranging the publication of this revised work by Junk Publishers).

For fish identifications and help with the taxonomical literature, I am indebted to Dr. Heraldo Britski (and also for permission to use the picture in Fig 5.13), Dr. Stanley Weitzman, Dr. Richard Vari, and Dr. William L. Fink. The Latin American Studies Center of the University of California, Los Angeles provided an initial grant that allowed me to gather a large body of literature on the taxonomy of South American fishes; its director, Dr. Johannes Wilbert, is also thanked for his support.

Juvenal Dácio and Jorge Dácio made the maps, graphs, and river profiles and Miss Karen Olsen executed the drawings. Mr. George Nakamura helped in preliminary editing and typed the manuscript.

Last, but not least, I say *obrigado* to the intrepid men who fished and collected data for me in the Rio Madeira valley: Sr. Dorval dos Santos, Sr. Valney Neves, Sr. Estácio Gomes, Sr. Raimundo Cécil Nascimento, Sr. Baixinho Cavalcante, Sr. Celestino Filho de Sousa, and Sr. Raimundo Deoclécio.

This work is dedicated to Charlie Bennett for stimulating my interest in the humid tropics.

Contents

Preface V

Acknowledgements VII

List of tables and figures XI

1. Physical and biological portrait of the Rio Madeira basin 1
    Geological history and river morphology 1
    Hydrochemistry 5
    River level fluctuation 10
    Vegetation 12
    Human modification of the Rio Madeira basin 16

2. Cultural backdrop of the Rio Madeira basin 17
    Exploration 17
    The Madeira-Mamoré Railway and the foundation of Porto Velho . . . 22
    Highways 23
    Population of the Rio Madeira basin 23
    Animal protein flow in the upper Rio Madeira valley 24
    Nature of fisheries in relation to urban centers 25

3. The fisheries 26
    Catfish fisheries 26
    Migratory characin fisheries 38
    Floodplain fisheries 43
    Flooded forest fisheries 51
    Weir fisheries 53
    Low water fishing in clearwater tributaries 56
    The new frontier: The Rio Mamoré and Rio Guaporé 56

4. Fishing area, effort, and yield 59
    Fisheries regions of the Rio Madeira drainage system 59

    Yields by species 59
    History of annual catches 60
    Seasonality in catches 63
    Catch per unit of effort 63

5. Natural history of the food fishes 69
    Fish migrations in the Rio Madeira basin 69
    The food chain sustaining the commercial fishes 72
    The food fishes 72
    Pimelodidae 73
    Doradidae 88
    Loricariidae 91
    Hypophthalmidae 91
    Characidae 91
    Prochilodontidae 100
    Curimatidae 105
    Anostomidae 106
    Hemiodontidae 106
    Erythrinidae 106
    Cichlidae 109
    Osteoglossidae 109
    Sciaenidae 113
    Clupeidae 113

6. Problems and prospects 117
    Relative productivity of Rio Madeira fisheries 117
    Direct management of fisheries 119
    Environmental protection 120

Bibliography 121

Author index 123

Index to scientific names 125

Subject index 127

List of tables and figures

Tables

1.1	Comparative measurements of salinity and suspended solids concentrations of Rio Madeira and other Amazonian rivers	10
3.1	The catfishes exploited at Teotônio and the months that each species migrates upstream and through the rapids	28
3.2	Migratory characins of the Rio Madeira	39
3.3	The common food fishes of the Rio Madeira floodplain and the months when they are most captured and the gear used to take each species	45
4.1	Total 1977–1979 catch of Rio Madeira food fishes	60

Figures

1.1	The Rio Madeira drainage system	2
1.2	Comparative discharges of Amazonian rivers	3
1.3	The upper Rio Madeira	3
1.4	The Teotônio cataract of the upper Rio Madeira	6
1.5	High alluvial bank characteristic of Rio Madeira	7
1.6	Profiles of the Rio Madeira and Rio Machado	8
1.7	Landsat satellite image of lower Rio Madeira and its confluence with the Rio Amazonas	9
1.8	Landsat satellite image of upper Rio Madeira	9
1.9	River level fluctuation of the Rio Madeira at Porto Velho between 1974 and 1979 and at Borba in 1979	11
1.10	Flooded forest of the Rio Madeira	13
1.11	The kapok cotton tree, the largest tree of Amazonian floodplains	14
1.12	Cassiterite, or tin-stone, mining in Rondônia	15
2.1	Drawing by Franz Keller of Rio Madeira fisherman at the cataracts	18
2.2	The highway network of the Amazon basin	20
2.3	An old locomotive of the Madeira-Mamoré Railway being conquered by the rainforest	21
2.4	Ferry used to cross a river along the Porto Velho/Manaus highway	21
3.1	The gaff used at the Teotônio rapids to catch large catfishes	30
3.2	Gaffing site at the Teotônio cataract of the Rio Madeira	30

3.3	*Candirú pintado* (*Pseudostegophilus* sp., Trichomycteridae)	31
3.4	*Candirú-açú* (*Cetopsis* sp., Cetopsidae)	31
3.5	Fisherman prepares to remove his handline hook from the stunned *jaú* (*Paulicea lutkeni*, Pimelodidae)	33
3.6	The *côvo*, or fish trap	33
3.7	The Ceará rock in the middle of the Teotônio cataract	34
3.8	Close-up view of Ceará rock	34
3.9	Fisherman pulling in a *jaú* (*Paulicea lutkeni*, Pimelodidae) with a castnet	35
3.10	The drifting deepwater-gillnet in the Rio Madeira channel	37
3.11	A. The drifting deepwater-gillnet used to catch catfishes near the bottom of the river. B. The river channel trotline	37
3.12	The 110 kg *piraíba* (*Brachyplatystoma filamentosum*, Pimelodidae)	40
3.13	The *rede de lance*, or seine, used in Rio Madeira fisheries	41
3.14	Schematic drawing of the capture of a school of spawning migratory characins descending the Rio Machado to breed in the Rio Madeira	41
3.15	The *zagáia*, or gig	46
3.16	The *pindá*-lure	46
3.17	Bow-and-arrow used by Rio Madeira fishermen	47
3.18	An archer fires an arrow at a *curimatá* (*Prochilodus nigricans*, Prochilodontidae) . . .	47
3.19	The *camurim* float	48
3.20	The harpoon used to kill the *pirarucu* (*Arapaima gigas*, Osteoglossidae)	49
3.21	The *curumim*-line used to catch *pirarucu*	50
3.22	The flooded forest trotline and gillnet	52
3.23	A fisherman constructing a flooded forest gillnet	54
3.24	A fish weir of the upper Rio Madeira	55
3.25	Drifting beach-gillnet used in clearwater tributaries of Rio Madeira	55
3.26	Crude fishing boat used by Bolivian fishermen	57
3.27	The *calhapo*, or underwater pen, used to transport live fishes	57
4.1	Total annual catches of the Porto Velho fishing fleet of the Rio Madeira	61
4.2	Total annual catches of the Teotônio cataract fisheries	61
4.3	Monthly catches of the Porto Velho fishing fleet in 1977, 1978 and 1979 in relation to water level	64
4.4	Mean man-day yield (kg) in years 1974–1979	65
4.5	Catch per unit of effort (CPUE) in terms of mean monthly kilogram yield per man-day fished for different types of fisheries of the Porto Velho fleet of the Rio Madeira in 1977	66
4.6	Catch per unit of effort (CPUE) in terms of kilogram yield per liter of fuel spent in 1977 . .	67
5.1	Fork length distribution of *dourada* (*Brachyplatystoma flavicans*, Pimelodidae)	74
5.2	The *dourada* (*Brachyplatystoma flavicans*, Pimelodidae)	75
5.3	The *filhote* (*Brachyplatystoma filamentosum*, Pimelodidae)	75
5.4	The *piramutaba* (*Brachyplatystoma vaillantii*, Pimelodidae)	77
5.5	The *babão* (*Goslinia platynema*, Pimelodidae)	77
5.6	The *bico de pato* (*Sorubim lima*, Pimelodidae)	79
5.7	The *mandi* (*Pimelodus blochii*, Pimelodidae)	79
5.8	The *jaú* (*Paulicea lutkeni*, Pimelodidae)	80
5.9	The *caparari* (*Pseudoplatystoma tigrinum*, Pimelodidae)	81
5.10	The *surubim* (*Pseudoplatystoma fasciatum*, Pimelodidae)	82
5.11	The *coroatá* (*Platynematichthys notatus*, Pimelodidae)	83
5.12	The *pirarara* (*Phractocephalus hemiliopterus*, Pimelodidae)	85

5.13   The *dourada zebra* (*Brachyplatystoma juruense*, Pimelodidae) 85
5.14   The *peixe lenha* (*Surubimichthys planiceps*, Pimelodidae) 86
5.15   The *dourada fita* (*Merodontotus tigrinus*, Pimelodidae). 86
5.16   The *barba-chata* (*Pinirampus pirinampu*, Pimelodidae) 87
5.17   The *pintadinho* or *piracatinga* (*Callophysus macropterus*, Pimelodidae) 88
5.18   The *cuiu-cuiu* (*Oxydoras niger*, Doradidae) 89
5.19   The *bacu comun* (*Pterodoras granulosus*, Doradidae) 90
5.20   The *bacu* or *bacu rebeca* (*Megaladoras irwini*, Doradidae) 90
5.21   The *bacu pedra* (*Lithodoras dorsalis*, Doradidae) 92
5.22   The *bodó* (*Plecostomus* sp., Loricariidae) 93
5.23   The *mapará* (*Hypophthalmus edentatus*, Hypophthalmidae) 94
5.24   The *jatuarana* (*Brycon* sp., Characidae) 94
5.25   Monthly distribution of annual catches of *jatuarana* by the Porto Velho fleet in the Rio Madeira in 1977, 1978 and 1979 95
5.26   The *matrinchão* (*Brycon* sp., Characidae) 96
5.27   The *tambaqui* (*Colossoma macropomum*, Characidae) 96
5.28   The *pirapitinga* (*Colossoma bidens*, Characidae) 99
5.29   A. The *pacu vermelho* (*Mylossoma albiscopus*, Characidae);   B. The *pacu branco* (*Mylossoma aureus*, Characidae) 99
5.30   The *pacu mafurá* (*Myleus* sp., Characidae) 101
5.31   A. The *sardinha comprida* (*Triportheus elongatus*, Characidae);   B. The *sardinha chata* (*Triportheus angulatus*, Characidae) 101
5.32   The *piranha caju* (*Serrasalmus nattereri*, Characidae) 102
5.33   The *curimatá* (*Prochilodus nigricans*, Prochilodontidae) 102
5.34   Monthly distribution of annual catches of *curimatá* by the Porto Velho fleet in the Rio Madeira in 1977, 1978 and 1979 104
5.35   Monthly distribution of annual catches of *jaraqui* by the Porto Velho fleet in the Rio Madeira in 1977, 1978 and 1979. 104
5.36   A. The *jaraqui escama grossa* (*Semaprochilodus theraponura*, Prochilodontidae);   B. The *jaraqui escama fina* (*Semaprochilodus taeniurus*) 105
5.37   A. The *branquinha cabeça lisa* (*Curimata altamazonica*, Curimatidae);   B. The *branquinha chora* (*Curimata latior*);   C. The *branquinha comun* (*Curimata vittata*);   D. The *cascudinha* or *chico duro* (*Curimata amazonica*). 107
5.38   A, B, E. Different colormorphs of the *aracu botafogo* (*Schizodon fasciatus*, Anostomidae);   C, D. Unidentified anostomids;   F. The *pião* (*Rhytiodus microlepis*, Anostomidae) 108
5.39   The *orana* (*Hemiodus* sp., Hemiodontidae) 110
5.40   The *traíra* (*Hoplias malabaricus*, Erythrinidae) 110
5.41   The *tucunaré* (*Cichla ocellaris*, Cichlidae) 111
5.42   The *cara-açu* (*Astronotus ocellatus*, Cichlidae) 111
5.43   Drawing of a *pirarucu* (*Arapaima gigas*, Osteoglossidae) from Franz Keller's *The Amazon and Madeira Rivers*. 112
5.44   The *aruanã* (*Osteoglossum bicirrhosum*, Osteoglossidae) 114
5.45   The *pescada* (*Plagioscion squamosissimus*, Sciaenidae) 115
5.46   The *apapá* (*Pellona castelnaeana*, Clupeidae). 115

CHAPTER 1

Physical and biological portrait of the Rio Madeira basin

The Rio Madeira, although overshadowed by the voluminous Rio Amazonas, of which it is a major tributary, is still one of the largest rivers in the world (Figs. 1.1 and 1.2). Its headwaters are born on the snow covered peaks of the Bolivian Andes, just east of La Paz, and flow over 1,500 km before reaching the Rio Amazonas. In the family of tributaries contributing to the Rio Solimões-Amazonas, it is unequalled in total drainage area, embracing about 1.3 million km^2, or nearly one-fifth of the entire Amazon basin. The Rio Madeira's total discharge is estimated to be about one trillion m^3 per year, or more than one-fifth of the total annual volume of the Rio Anazonas; the massive Rio Negro, however, has a larger annual discharge which is estimated to be on the order of 1.4 trillion m^3 per year (Hidrologia 1975-1979). The Rio Madeira's annual volume is about twice that of the Mississippi's and nearly equal to that of Africa's largest river, the Zaire (Fig. 1.2).

With the Rio Guaporé added, the Rio Madeira has the longest south to north trending course of any river lying entirely within the tropics. For the South American continent it has been a major dispersal and exchange route for aquatic animals in the distant La Plata and Amazon basins. Even today, during the annual floods, the headwaters of the Rio Guaporé are met by the expanding Pantanal, the huge swamp that drains into the Paraguay-La Plata system. The fish faunas of the La Plata and Amazon basins are very similar, and this is strong zoogeographical evidence of past hydrographical links between these large drainage systems (Pearson 1937; Menezes 1970).

For most of its length the Rio Madeira flows along the western flank of the Brazilian Shield. It has no large affluents originating in the Amazon Lowlands, and indeed the tributaries of the Rio Purus to the west, reach almost to its left bank in some cases. The large tributaries of the Rio Madeira are born in the Andes or on the Brazilian Shield and, as will be seen later, these two geological formations largely control the hydrochemistry of the system.

Geological history and river morphology

The Rio Madeira appears to have assumed its present course in the Quaternary Period, or no more than about three million years ago. During most of the early Tertiary Period (about 15-65 million years ago), and before the rise of the Andes, the rivers of the western Brazilian Shield probably drained northwest and into the western Amazon basin, and perhaps there joined a larger river that flowed into the Pacific at about the Gulf of Guayaquil. With Andean orogeny in the Miocene and after, the Pacific outlet was cut off and, because the Guiana and Brazilian Shields to the east were still joined, a huge lake or series of lakes is hypothesized to have formed in the Amazon Lowlands. Sometime in the Pliocene or early Pleistocene, the future Rio Amazonas cut through the low-lying area

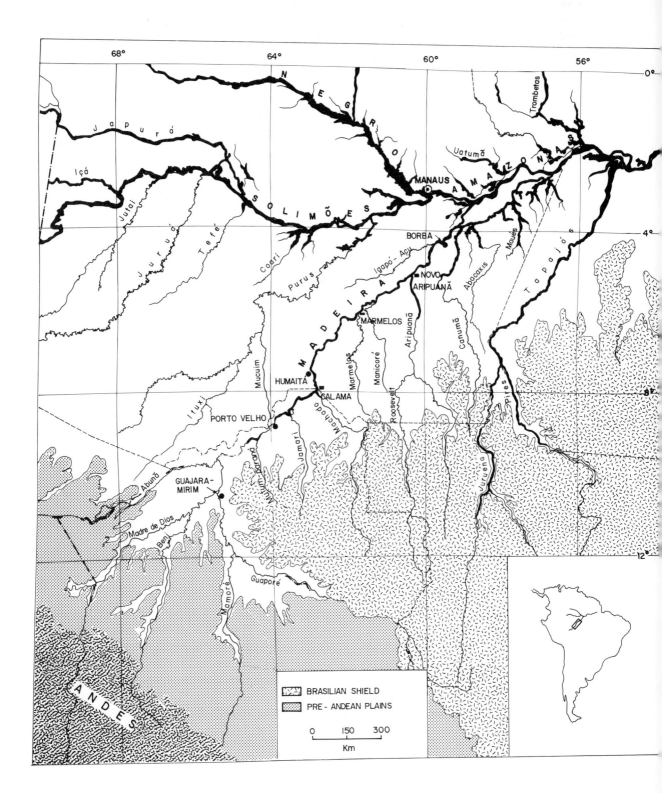

Fig. 1.1 The Rio Madeira drainage system.

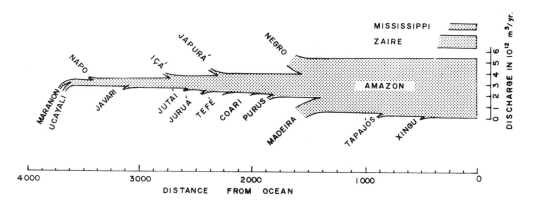

Fig. 1.2 Comparative discharges of Amazonian rivers. After Gibbs (1967).

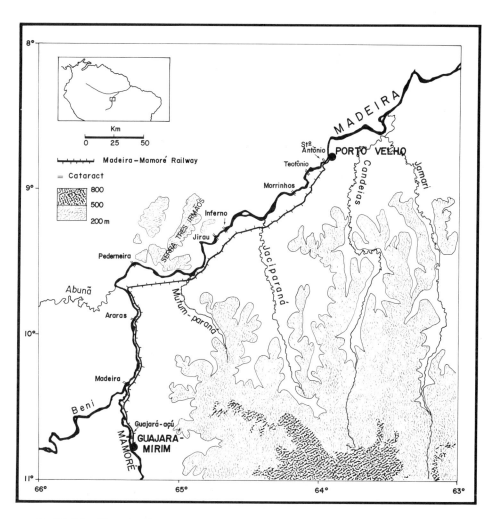

Fig. 1.3 The upper Rio Madeira. Note that the cataracts begin above Porto Velho. The Madeira-Mamoré Railway is also shown. The straight stretch of track running west from the Mutum-paraná cuts across an extremely level area built of lacustrine sediments. Before the eastern Bolivian rivers cut through the low-lying area between the Serra Tres Irmãos and the main part of the Brazilian Shield, a lake was formed west of the Mutum-paraná.

between the respective Shield areas and the Amazon basin began to drain into the Atlantic Ocean (Beurlen 1970; Fittkau 1974; Grabert 1967).

It is unclear whether the first appearance of the Rio Madeira was coeval with that of the Rio Amazonas, or if the large tributary, as seen today, assumed its present course only when the eastern Bolivian rivers began to drain into the Amazon, which is thought to have been in the Quaternary (Grabert 1967). With the rise of the Andes, eastern Bolivian drainage was obstructed not only to the west by the new mountain chain, but also to the north, south, and west by hills and highlands of diverse geological origin. Lacustrine deposits in eastern Bolivia and southeast of the mouth of the Rio Beni in Brazil, lend ample testimony to the existence of an ancient lake or lakes. Sometime in the early Quaternary, the low-lying area between the present day Serra Tres Irmãos and the main part of the western Brazilian Shield was ruptured, the eastern Bolivian lake was drained, and the Andean rivers began to flow into what is now the Rio Madeira (Fig. 1.3). The gap that was cut through the western most extension of the Brazilian Shield, now demarcated by the Madeira rapids, is still unable to accomodate the huge volumes of water that are poured into it each year with the floods, and consequently the eastern Bolivian rivers centering on the Rio Beni and Rio Mamoré are partially backed up and overflow onto the low adjacent plains. Devenan (1966) has given an excellent description of these savanna regions in relation to ridge field farming by early Amerinds.

The 356 km stretch of river between Porto Velho on the Rio Madeira and Guajará-Mirim on the Rio Mamoré cascades through about 12 to 20 cataracts – depending on what you recognize as a cataract – formed mostly of resistant granitic rocks (Fig 1.3). The Rio Madeira has not yet had time to escavate its riverbed, and the cataracts are the result. There are no true falls in the Madeira rapids, and overall the individual cataracts are rather short – no more than about one kilometer in length – and have modest declivities. The largest of the cataracts is the Cachoeira do Teotônio, which lies about 20 km upriver of Porto Velho. At Teotônio a rocky bulwark stretches across the entire width of the river, and causes an eight meter drop between smooth water above and below the rapids. During the lowest water period, about three-fourths of the granitic bulwark is emerged and the violent channel flows close to the left bank (Fig. 1.4). The torrential waters ushered through the Teotônio cataract present an obstacle to upstream migrating fishes and, as will be discussed later, this is taken advantage of by commercial fishermen.

The Rio Madeira is about 900 km in length between its mouth and the first cataracts above Porto Velho, and for most of this distance it is confined by high alluvial levees or terra firme banks (Fig 1.5). The floodplain is relatively small and probably accounts for no more than about 1,000 km^2. This may be compared to the Rio Solimões-Amazonas which has an estimated 15,000 km^2 of floodplain and the Rio Purus with about 2,500 km^2 (Gourou 1950; Pires 1974). Relative to total discharge, the Rio Solimões and Rio Purus have at least seven times more flooded forest than the Rio Madeira. The floodplain that does exist along the Rio Madeira is also high for the most part, and this is perhaps a reflection of both underlying topography and the heavy load of sediment that the river carries, which historically, has filled up the adjacent low-lying river valley areas. Annual flooding time of inundation forest along the middle Rio Amazonas has been reported to be about six months (Smith 1979), while the average for the Rio Madeira (above the Rio Aripuanã) is closer to two to three months. The relatively lower lying floodplains of the Rio Madeira affluents, however, have longer inundation periods than the principal river and are comparable to the middle Rio Amazonas, or about six months (Fig. 1.6). The Rio Madeira floodplain is largest near its confluence with the Rio Amazonas (Figs. 1.7 and 1.8), and flooding times are also longer than in the middle and upper course of the river. The longer flooding times are the result of differential water levels between the Rio Madeira and Rio Amazonas. The Rio Madeira reaches its flood peak in March or early April, whereas the Rio Amazonas attains its high in June or July, and thus the tributary is dammed back by the principal river.

The large rightbank tributaries rise at 600-1,000 m on the Brazilian Shield. Most of their lengths are interrupted by cataracts that make boat travel into the headwaters extremely difficult and, in fact, greatly hindered the early exploration of this region. The last cataracts are located where the Brazilian Shield meets the Amazon Lowlands, and this is usually in the inferior courses of these affluents. Above the first cataracts of these tributaries, the river channel is well defined and there is little floodplain; where the Amazon lowlands are met, the floodplains of the larger rivers average about 500-1000 m in width on at least one side, and are mostly covered by flooded forest. There are also small lakes accompanying the lower courses of these rivers.

Hydrochemistry

Surface geology largely determines hydrochemistry in natural ecosystems where population densities are low and there has been no industrial or agricultural pollution. This is the case for most of the Amazonian drainage system. Amazonian rivers drain three main geological formations, namely the Andes, the Guiana and Brazilian Shields, and the rainforest covered lowlands. Each of these formations produces one or two general types of streams or rivers that can be recognized hydrochemically, and even visually in most cases. The affluents of the Rio Madeira tap the main geological trio, though the hydrochemistry of the procicipal river is largely controlled by nutrient enriched water draining the western Andean slopes.

Both complex mixture of rock types, and high elevation that greatly increases physical weathering, account for the fact that the Andes play the role of nutrient bank for the Rio Madeira, as indeed for much of the Amazon Basin (Gibbs 1967; Sioli 1967, 1968). The streams and rivers emanating from the Andean Cordillera are heavily charged with suspended materials picked up from erosion, and render the rivers downstream of them turbid; these are often referred to as whitewater rivers, both regionally and in the literature, though they would best be described as colored cafe-au-lait. The Andean tributaries flowing into the Rio Madeira meander through the east Bolivian savannas, and indeed greatly inundate them during the flooding season, but there are no studies indicating what effect this might have on the hydrochemistry of the Rio Madeira. As will be discussed later, it appears that the swampy savannas may provide, to a certain extent, a nutrient trap for the waters injected into them each year with the floods; these nutrients, spread out in a greatly expanded aquatic environment for a few months, may be a key factor in building up a food chain, though phytoplankton, on which many young and adult fish are sustained. Some catfish species from the Rio Madeira appear to migrate upstream and into the savanna regions to take advantage of this.

Hydrochemically the Rio Madeira is similar to the Rio Solimões-Amazonas, but there are some differences. Both rivers owe their high suspended loads and turbidity to Andean materials and, as mentioned, most of their nutrients come from the cordilleran slopes. At the ecosystem level, the Rio Madeira is much less influenced by streams and rivers rising in the Amazon Lowlands than is the Rio Solimões-Amazonas, though the overall role of these rainforest affluents, especially the blackwater, highly acidic ones, is still unknown. Gibbs (1967) found that the Rio Madeira has a significantly higher level of material in suspension than does the Rio Solimões-Amazonas, and also a slightly higher overall salinity (Table 1.1). The Rio Madeira, however, has a much smaller floodplain than the Rio Solimões-Amazonas, and thus its salts are to much less avail in total primary production.

For most of the year the Rio Madeira has a very low Secchi Disk transparency (SDT)*, being no more than a few centimeters because of the high turbidity of the water. Unlike the Rio Solimões-Amazonas, however, the Rio Madeira clarifies to about one meter SDT for a short period during the low water period in most years. The hydrological data for the 1974-1979 period suggest that if river level remains above about the four meter mark on the Porto Velho staff gauge, then the Rio Madeira does not clarify (Fig 1.8); this was the case in 1977.

*A Secchi Disk is a round, white disk that is used to measure water transparency by lowering it into the water with a cord. The depth at which the disk disappears from sight is taken to be the transparency.

Fig 1.4 The Teotônio cataract (Cachoeira do Teotônio) of the upper Rio Madeira.

Fig. 1.5 High alluvial bank characteristic of Rio Madeira. Note that the floodplain is forested. Subsequent to the floods, high alluvial banks often cave in (*terras caídas*) and form woody shore areas, a favorite habitat of some fish species during the low water period.

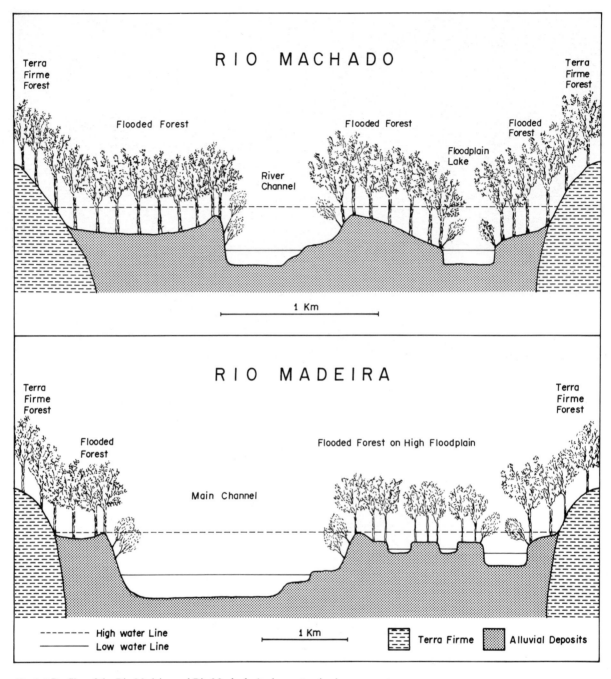

Fig. 1.6 Profiles of the Rio Madeira and Rio Machado (a clearwater river).

The turbidity of the Rio Madeira is undoubtedly a function of rainfall and erosion in the Andes. It appears that in most years Andean erosion during the 'dry season' is reduced sufficiently, at least for a few weeks, to allow the Rio Madeira to clarify.

In stark contrast to the Rio Madeira, are its non-Andean tributaries, most of which drain the western flank of the Brazilian Shield, and to a much lesser extent, the Amazon Lowlands. These rivers have minimum suspended loads and are referred to

Fig. 1.7 Landsat satellite image of lower Rio Madeira, and its confluence with the Rio Amazonas. The photograph was taken in Band 7 on 30 July 1977.

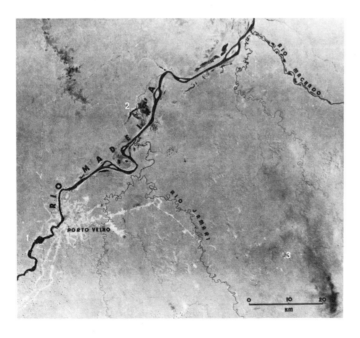

Fig. 1.8 Landsat satellite image of upper Rio Madeira. Note the small floodplain in comparison to the lower Rio Madeira and its confluence with the Rio Amazonas. 1) Cachoeira do Teotônio; 2) Cuniã, the largest floodplain area of the upper Rio Madeira; 3) general area of tinstone (cassiterite) mining. The photograph was taken in Band 7 on 17 May 1967.

Table 1.1 Comparative measurements of salinity and suspended solids concentrations of Rio Madeira and other Amazonian rivers. From Gibbs (1967).

River	Salinity		Suspended solids concentrations	
	Low water (ppm)	High water (ppm)	Low water (ppm)	High water (ppm)
Rio Madeira	68	50	15	359
Amazonas	48	28	22	123
Negro	6	4	1	9
Tapajós	11	6	1	4

locally as clearwaters of blackwaters. All of the large right bank tributaries of the Rio Madeira, including the Canumã, Aripuanã, Marmelos, Manicoré, Machado, and Jamari, flow off the Brazilian Shield and their waters are clear when looked at through a glass. Secchi Disk transparencies range from about 1.5-4 m, and conductivity readings are generally below 20 uS_20, which is to say very low. The absence in these waterbodies of intensive phytoplankton blooms and aquatic macrophytes, or herbaceous vegetation, strongly suggest that nutrient levels are minimal, as also indicated by low conductivities.

Rivers that are tinted by humic acids, the large Rio Negro being the most famous, are referred to in the Amazon as blackwater rivers because of their tealike color. The Rio Madeira has only one large blackwater tributary, and this is the Rio Preto do Igapó-Açu found in the lower course. The humic acids responsible for the blackness of the Rio Negro have been linked to sandy soils supporting a special, low vegetation type called *caatinga* (Klinge 1967). The acidity of the granitic and gneissic soils inhibits the breakdown of the humic acids found in the dead vegetation, and hence they are leached into the rivers. It does not appear that the blackwater of the Rio Preto do Igapó-Açu can be linked to sandy soils and *caatinga* vegetation, and there are other blackwater streams and waterbodies, especially smaller ones found along or adjacent to the Rio Madeira floodplain, which appear to originate in high rainforest with latosolic soils. Many of the floodplain lakes of the Rio Madeira are filled with blackwater, whose sources must lie close on the adjacent terra firme.

River level fluctuation

The Rio Madeira drainage system lies in the tropical zone, stretching from about the twentieth parallel of latitude in the southern hemisphere to near the equator. Rainfall is seasonal, and this is most saliently expressed in the landscape with the rising and falling of the large rivers. The solar equator, or the latitude where the sun's rays are mostly directly overhead at any particular time of the year, is located for six months in the tropical zone of each hemisphere. In the southern hemisphere the sun's rays are most directly overhead between September and March, and this period embraces most of the high precipitation months. Large rivers take time to fill and drain, and thus do not always reflect, in their monthly oscillations, the menstrual distribution of rainfall. Water level data for the Rio Madeira show that it rises between the September and March equinoxes and falls in the complementary period. The river, then, rises for about six months and falls for about six months. Allowing of course for the period needed for the river to fill after the beginning of rising water, and to fall subsequent to the peak of the annual flood, the high water season of the Rio Madeira is between about December and June. This period also correlates well with the inundation times of the flooded forests found on the floodplains.

The average annual fluctuation of the Rio Madeira at Porto Velho in the period 1974-1979 was about 12.4 m, while near Borba in the lower course of the river it was 10.5 m (Hidrologia 1974–1979), or about 2 m less (Fig 1.9). This may be compared to the oscillation of the middle Rio Amazonas that is about 7 m (Smith, 1979), and the lower Rio Negro – which is dammed back by the Rio Solimões-Amazonas – that averages around 10 m (Portobras 1976-1979). The lower Rio Madeira reaches its peak flood level about one month after that of the upper course of the river. This is due in

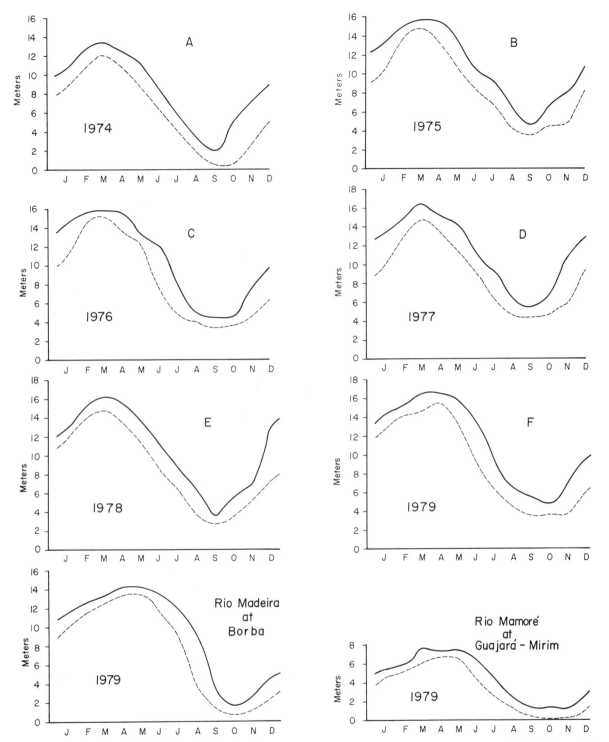

Fig. 1.9 River level fluctuation of the Rio Madeira at Porto Velho between 1974 and 1979 (A through D) and at Borba in 1979. Porto Velho is in the upper course of the Rio Madeira, while Borba is near the mouth. Note the extremely low water period in 1974 as indicated by the Porto Velho hydrograph lettered A. The peak of the Rio Madeira flood is reached later in the lower than the upper course, because the Rio Amazonas dams back its large tributary. Also included is the river level fluctuation of the Rio Mamoré at Guajará-Mirim in 1979; the Rio Madeira rapids are unable to accomodate the huge volumes of water poured over them during the floods, and thus the Rio Mamoré and Rio Beni are partially dammed back and have an extended flooding season. Water level data from Hidrologia (1974-1979).

large part to the fact that the Rio Amazonas peaks several months after the high water level of most of the Rio Madeira, and thus partially dams back the lower course of the large tributary.

There is a great regularity in the seasonal flow of the Rio Madeira. In the period 1974-1979, the flood peak was reached between 08 March and 03 April, and in three of these years (1974-1976) it fell on the same or almost the same date (17 or 18 March). The minimum low water level for the same period ranged between 05 September and 01 October. The river level data available for the Rio Madeira suggest that extremely small floods are followed by extremely low water periods, as was the case in 1974 (Fig. 1.9). The peak range between maximum and minimum floods in the period 1974-1979 was about two meters, while there was at least a four meter difference between the extremely low water period of 1974 and the minimum reached in 1977. This indicates that there is less variation in flood than in low water levels.

The peak of the flood of the Rio Mamoré at Guajará-Mirim is about a month after that of the Rio Madeira at Porto Velho, and likewise, the minimum low water level is also reached later (Fig. 1.9). The flattened peak of the Rio Mamoré flood is due to the inability of the Rio Madeira, in its cataract stretch, to accomodate the huge volumes of water that its upper tributaries – along with Rio Mamoré, the Rio Beni and Rio Abunã – pour into it. As mentioned earlier, the tributaries are backed up and inundate their low floodplains. The large right bank tributaries of the Rio Madeira have about the same flooding regime as the main river, consequently are not invaded by turbid water.

Vegetation

The watershed of the Rio Madeira drainage basin is covered with a diverse range of vegetation types, including cloud forest on the Andean slopes, savannas in the eastern Bolivian Plains, shrubby communities on the higher parts of the Brazilian Shield, and tropical rainforest in most of the Amazonian Lowlands. These general vegetation types need not be discussed here – other than to point out that vegetation is important in protecting watersheds from serious erosion – but of more immediate relevance to the Rio Madeira fisheries are flooded forests that cover the floodplains and aquatic herbaceous plants found mostly along the Rio Madeira or in open waters of its floodplain.

In the Amazon, flooded forests are usually called *igapó* (Fig. 1.10). Biogeographical studies of flooded forests have only just now begun, but it is clear that there are significant floristic differences between at least some blackwater and turbid river floodplains, while the clearwater flooded forests may be somewhat transitional between the other two general types (Prance 1978). The flooded forests of the Rio Madeira and its large right bank tributaries are characterized by high forest, with the kapok cotton tree (*Ceiba pentandra*, Bombacaceae) reaching at least 40 m in height (Fig. 1.11). The flooded forest is adapted to withstand long periods of flooding, and indeed some of the seedlings and shore shrubs are totally or partially submerged for as long as ten months each year. As a community, flooded forests fruit mostly during the high water season, and this may be an adaptation for seed dispersal by both water and fishes (Goulding 1980). The flooded forests play a special role in Amazon fisheries ecology as they supply much if not most of the food – fruits, seeds, detritus, and insects – on which many important commercial species depend for their nutrition.

The distribution of aquatic macrophytes, or herbaceous plants, is restricted mostly to the Rio Madeira floodplain where there is a annual injection of nutrients from the turbid water to support them. As communities they are called 'floating meadows' because most of the vegetation types float; those species, such as grasses of the genera *Paspalum* and *Echinochloa*, that are initially grounded in the substrate, eventually become detached and float as the flood level rises (Junk 1970). The root zones of floating meadows are known to support incredible numbers of arthropods (Junk 1973), and the habitat is probably important in food chains leading to fishes, especially alevins.

Fig. 1.10 Flooded forest of the Rio Madeira. The surface of the water is covered with a carpet of herbaceous plants that were transported from open waterbodies with the inundation of the surrounding floodplain forest. Water depth is about 8 m.

Fig. 1.11 The kapok cotton tree (*Ceiba pentandra*, Bombacaceae), the largest tree of Amazonian floodplains. A. View of the forested floodplain of the Rio Machado, with the kapok cotton tree towering over all others. B. The massive, buttressed trunk of the kapok cotton tree.

Human modification of the Rio Madeira basin

Human economy has now reached the point in the Rio Madeira basin of seriously modifying ecosystems. Deforestation, river impoundment, and mining will all take a heavy toll on natural ecosystems that are unfortuately very poorly understood in this region.

Deforestation is spreading mostly from the new highway network that cuts across Rondônia. The Porto Velho/Cuiabá highway bisects Rondonia and, of great importance to the aquatic ecosystems, it traverses the watersheds of several of the large right bank tributaries. Within a few years the upper watersheds of the Rio Jamari, Rio Machado, and Rio Aripuanã will be largely deforested. This will probably lead to increased erosion and greater turbidity of the clearwater rivers. There may be many fish species, especially smaller forms, that are adapted only to live in relatively clearwaters, and these will certainly be affected by any serious modification of water transparency. Other serious effects of erosion on the equatic ecosystems are unclear at this time.

The Rio Madeira floodplain, though relatively small, has been little deforested, though selective logging operations have increased considerably in the past few years. Since 1976, for example, I have noticed the disappearance of many of the larger trees in the floodplains of the Rio Madeira and Rio Machado. Giants such as the kapok cotton tree *(Ceiba pentandra)* are now being extensively cut for timber. Roadwork has already begun to connect villages along the upper and middle Rio Madeira with the main highway arteries, including the Porto Velho/Manaus and Transamazon highways. Once these roads are opened up, deforestation of the Rio Madeira floodplain will become almost inevitable.

Accompanying deforestation and the development of cattle ranching and crop agriculture will also be the mining industry. Cassiterite, or tin-stone, mining in a stretch reaching from the Rio Jamari to the headwaters of the Rio Aripuanã, is already, as of 1980, polluting some of the rivers of the region. Cassiterite is removed from alluvial, eluvial, or colluvial deposits, usually found near streams, by placer mining (Fig. 1.12). The dirt and gravel are washed in hydrological sluices, and the tailing is often deposited in streams to be carried away. This leads to stream and river pollution. The Rio Jacundá and the Rio Preto, tributaries of the Rio Machado, appeared to be the most seriously polluted rivers in Rondônia in 1980. Normally clearwater rivers, they were discolored by the mining waste. The ecological effects of this pollution on the streams and rivers has not been studied.

In 1980 there were no large dams in the Rio Madeira basin, though all of the major tributaries were being intensively surveyed for future impoundment. The clearwater rivers, with their series of cataracts and low suspended loads, offer auspicious conditions for dam sites. Planners hope that hydroelectric power can soon replace the expensive thermoelectric plants that are at present supplying the region's energy. The first dam in the Rio Madeira basin is planned for the Rio Jamari at the Samoel Cataract, about 60 km from Porto Velho. Completion of the dam is planned for 1984 or 1985. As population continues to grow along the Porto Velho/Cuiabá highway and frontier cities such as Jiparaná and Vilhena develop more fully, the Rio Machado and Rio Aripuanã will probably be dammed to supply the badly needed energy for this region. Brazilian and Bolivian government officials have also discussed the possibility of damming the Rio Madeira at some point in its cataract stretch, and both of the respective countries would be supplied with energy from this huge dam (Anonymous 1977). The Rio Madeira, however, has a very heavy silt load and transports large quantities of wood, and would present many technological problems that the clearwater affluents would not.

Fig. 1.12 Cassiterite, or tin-stone, mining in Rondônia.

CHAPTER 2

Cultural backdrop of the Rio Madeira basin

Exploration

The Rio Madeira basin was not an easy region to explore. There appear to have been five factors that mitigated against its early exploration, namely, its length and the length of its tributaries, cataracts, geographical isolation from the Atlantic littoral, malaria, and probably – though the historical evidence is not conclusive – a relative small Amerind population that would have attracted, as it were, more Portuguese slavers and traders. The first significant document that we have of a major expedition into the Rio Madeira basin dates to the mid-seventeenth century when the Portuguese trailblazer (*bandeirante*), Antonio Tavares Raposo, under the auspices of the Portuguese Crown, led his men from São Paulo to the Rio Guaporé, and then down the Rio Mamoré and Rio Madeira and on to the mouth of the Rio Amazonas (Cortesão 1958). The Tavares expedition was charged with increasing Portuguese territory in the *terra incognita* through which ran the Line of Tordesillas dividing South America between the Portuguese and Spanish empires. Unfortunately the Tavares expedition did not have much of a chronicler along, and thus history is robbed of the first glimpse of the Rio Madeira before the Amerinds were largely destroyed by disease and enslavement in the subsequent conquest of the region.

In the eighteenth century, official expeditions destined to explore the Rio Madeira were dispatched from Belém do Pará at the mouth of the Amazon river. The *relatórios*, or reports, of the period make it clear that the cataract stretch of the Rio Madeira took about as much time (45 and 53 days cited by two expeditions) to navigate as did the entire stretch of river from Belém to the rapids (Abreu 1930; Ferreira 1959; Hugo 1959). Nevertheless, by the end of the eighteenth century, there was a steady traffic bringing gold from the placers in Mato Grosso, and carrying merchandise upstream in return (Ferreira 1957). By 1800 the government had located permanent porters at the cataracts in an attempt to improve transportation. The cataracts had also been given the names by this time that they conserve to this day.

The first 'scientific' expedition to be led into the Rio Madeira basin was headed by Alexandre Rodrigues Ferreira (Ferreira 1972). He ascended the Rio Madeira in 1789, and reported to have lost 62 Indians that died in route. Ferreira left a series of *relatórios*, drawn mostly from his experiences over a large part of the Amazon basin, but did not give a coherent picture of the Rio Madeira as a system. One of his *relatórios* presents an important list of the vernacular names of the fishes he found; this document shows that the vulgar names of most of the common fishes were already established by this time. The Portuguese were little prepared to christen a fauna as diverse as the Amazon one, and either adopted or bastardized Amerind names for most of the piscine fauna.

In the nineteenth century, the Amazon would be visited by what are now the illustrious names of

Fig. 2.1 Drawing by Franz Keller of Rio Madeira fisherman at the cataracts. From Keller (1874).

naturalists like Spix, Martius, Wallace, Bates, Spruce, and Aggasiz. The Rio Madeira, however, had two other great river systems – the Rio Solimões-Amazonas and Rio Negro – competing with it for the attention of these naturalists. No more than the mouth area of the Rio Madeira would be reconnaissanced by the great nineteenth century naturalists, and thus, as was largely the case up to then, the large turbid water tributary would continue to flow in literary darkness.

By the 1860's both the Bolivian and Brazilian governments were warming to the idea of a transportation system that would circumvent the Madeira rapids. Franz Keller, an engineer, and his father were employed by the Brazilian government to explore the Madeira rapids and to suggest the most practical means by which a better transportation system could be effected. Their explorations were confined to near the river and not very detailed from an engineering standpoint. The Kellers suggested three alternatives that were already in the air, namely, the escavation of a canal, using pullies at the individual cataracts, and perhaps the best they thought, the construction of a railway to surpass the rapids. Franz Keller, certainly a better tourist and geographer than engineer, fortunately left a handsome volume discussing his travels and the natural history of the region. He discussed the vegetation, hunting and fishing (Fig. 2.1), hydrography, Amerinds, and other subjects of interest (Keller 1874). Even today Keller's book remains the only general treatment of the natural history of the Rio Madeira.

In the early twentieth century, two names – other than a few lesser ones associated with the construction of the railway – are most relevant to the exploration of the Rio Madeira basin. For more than three decades, beginning in the 1880's, Candido Mariano da Silva Rondon, better known simply as Coronel Rondon, explored the southern highlands of the Amazon and layed more than 3,000 km of telegraph line, thus connecting this region telegraphically to the rest of Brazil. Rondon himself did not leave a large natural history account of his experiences, but he did much to help scientists advance field research in the interior of Brazil; such illustrious names in Brazilian science as F.C. Hoehne, Alípio Miranda Ribeiro, Adolpho Ducke, and H. von Ihering were given their best opportunities to do research by accompanying Rondon on his expeditions. Rondon also sent many Amerind artifacts and plant and animal specimens to the Museu Nacional (Rio de Janeiro), and his collections accounted for the major contribution in the one century of existence of that institution (Ribeiro, 1976).

In 1913, Teddy Roosevelt, still stinging from his unsuccessful attempt to win the presidency of the United States for a third term, resolved to do a lecture tour in South America, but more audaciosly in tune with his adventurous spirit, decided also to combine his lectures with an exploration trip. The Foreign Minister of Brazil suggested that Roosevelt join forces with their foremost explorer, Coronel Rondon, and that together they should explore the Rio da Dúvida, or River of Doubt, whose headwaters had been discovered by Rondon himself, but at the time it could not be determined whether the river was an affluent of the Tapajós, Madeira, or the Jiparaná (Machado), an affluent of the Madeira. The Roosevelt-Rondon Expedition was vividly described in the ex-president's book, *Through the Brazilian Wilderness* (Roosevelt 1914). Roosevelt gave a reasonable description of the nature of a river flowing off the Brazilian Shield and into the Amazon Lowlands. Roosevelt's book is also important in Amazon ichthyology, as was pointed out by Myers (1949), in that more than any other work it brought the *piranha* to the attention of the world. The *piranha* subsequently has become as famous as Roosevelt himself.

There are no longer any 'doubtful' rivers in the Rio Madeira basin, but about all that we know about some of them is their geographical coordinates. Aerial photographs, especially the Randam series shot at a scale of 1:125,000 and the Landsat satellite images shot in seven wave lengths, have given us an impressive view from above of the Amazon system. More importantly, the Landsat satellite passes over any particular area every eighteen days, and thus its images can be used to monitor major vegetation disturbances.

Even today the Rio Madeira has not yet broken out of its geographical isolation. It is interesting, if

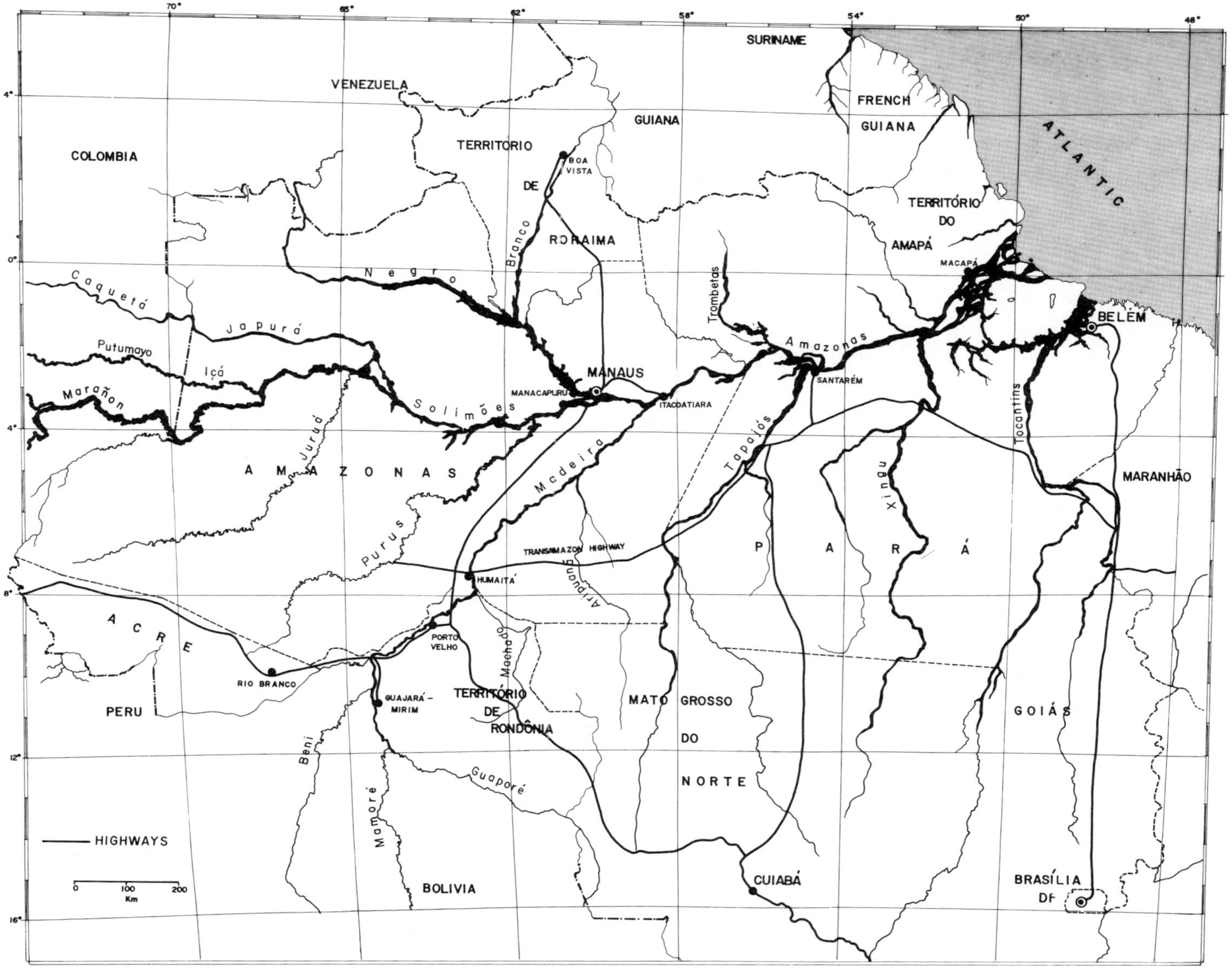

Fig. 2.2 The highway network of the Amazon basin.

Fig. 2.3 An old locomotive of the Madeira-Mamoré Railway being conquered by the rainforest. The picture was taken in 1980.

Fig. 2.4 Ferries, such as the one above, are used to cross the rivers along the Porto Velho/Manaus highway.

not ironic, to note that transportation costs and problems still hinder the proper scientific exploration of the Rio Madeira basin. Today it is not the cataracts, but the high price of fuel, that makes the exploration of the Rio Madeira and its large tributaries too expensive to attract very many people. That will probably continue to be true for some time to come.

The Madeira-Mamoré Railway and the foundation of Porto Velho

The construction of a rainforest railway connecting the Rio Madeira and Rio Mamoré between their intervening cataracts, a turbulent stretch of about 360 km, is undoubtedly the greatest spectacle in Amazon transportation history, even vis-à-vis the present highway network. The Madeira-Mamoré Railway, as it would be called, layed the tracks, so to speak, for the future development of the Rio Madeira basin (see Figs. 1.3 and 2.3).

The idea of a railway to surpass the Madeira rapids was discussed simultaneously in the 1860's in both Brazil and Bolivia, but its implementation would wait for the likes of Colonel George Church, an American of diverse interests, including soldiering, engineering, anthropology, geography, and the least of which was not, making money. With consent from the respective governments for the construction of a railway, Church obtained English financing and the first attempt at construction was begun in 1871, but abandoned by 1874 because of poor working conditions and a high mortality rate due mostly to malaria and dysentery. The first attempt was followed by a second with a fresh contract with P.T. Collins, an American engineering concern (Craig 1907). A high fatality rate and pay problems forced the Americans to leave in 1879, and further Brazilian efforts, until the Treaty of Petrópolis in 1903, withnessed very little significant work on completing the railway.

In the Treaty of Petrópolis, Brazil agreed with Bolivia to construct the railway in exchange for concessions of land in the Acre Territory in the headwaters of the Rio Purus and Rio Juruá, an area that was known to be rich in rubber trees and where a large number of immigrants from the state of Ceará had already settled (Bowman 1913). Serious work on the railway began anew in 1907 and a leading firm of American railway engineers was employed; the control of the railway passed in effect to the American magnate, Percival Farquar, whom, at the time, had huge financial assets, both in railroads and other businesses, in Brazil. Workers were scoured from the United States, the Caribbean, Europe, and much of Brazil. In 1912, the 360 km of railway were completed. The Madeira-Mamoré Railway was a costly venture, both in terms of construction costs and human lives. An estimated 6,000 men died during the construction of the rainforest railway (Ferreira 1957).

The Madeira-Mamoré was founded on the financial hope that rubber latex prices would remain high and that this commodity would be its main freight. The end of the rubber boom, however, coincided with the completion of the railway; in fact, 1912 and 1913 were the two most financially successful years in the history of the venture (Ferreira 1957). The drastic drop in rubber prices – due to competition from East Asian plantations – foreclosed whatever transportation advantages the Madeira-Mamoré Railway might have offered at the time. In 1931, control of the railway was wrested from Farquar by the Brazilian government, and it was financed, always at a defecit, until its death in 1965.

The terminus of the railway below the Madeira cataracts gave birth to Porto Velho, which is not at all, as its name implies, an old port, but dates from about 1913. An English journalist who visited the railway town at the time described it like this. 'Porto Velho had a population of about three hundred. There were Americans, Germans, English, Brazilians, a few Frenchmen, Portuguese, some Spaniards, and a crowd of negroes and negresses' (Tomlinson 1928). Porto Velho grew slowly until the 1960's, but the opening of the new highways set the stage for large scale migration to the region and the city. Porto Velho is perhaps now the bustling economic center of the region that railway planners once hoped it would be.

Highways

Until the mid-1960's, most of the Rio Madeira basin was isolated geographically and transportation was limited mostly to river travel. The boat trip from Manaus to Porto Velho took five or six days, and even longer from Belém. The Madeira-Mamoré Railway, as discussed above, did the region very little good as there were no valuable products that it could bring to port, and, in the end, one still had to travel by boat to make contact with other centers in the Amazon or most of the rest of Brazil. Limited air traffic helped to some extent in the 1950's and 1960's, but became of much more significance only when Porto Velho grew in the 1970's and jet service was begun.

The construction of the Cuiabá/Porto Velho Highway across the western flank of the Brazilian Shield (Fig. 2.2), largely following the telegraph line layed by Coronel Rondon, broke the isolation of the upper Rio Madeira basin and set the stage for large scale immigration from the states of Rio Grande do Sul, Paraná, Santa Catarina, Espírito Santo, and Mato Grosso (IBGE 1975, 1979; Thery 1976; Wesche 1978). Most of the 1,500 km long highway is unpaved and is shrouded in dust during the dry season or deeply rutted during the wet season when trucks and cars get mired in mud. Nevertheless, it is now one of the lifelines connecting the growing population of Rondônia with Central, Southeastern, and Southern Brazil.

In 1973, the Porto Velho/Manaus Highway was opened and is now paved but requires five ferries to complete the trip (Fig. 2.4). The 900 km long highway follows the low divide between the Rio Purus and Rio Madeira. The Transamazon intersects the Porto Velho/Manaus Highway near the city of Humaitá on the left bank of the Rio Madeira. Because of its geographical situation, Porto Velho has assumed the role of economic hub of the southwestern Amazon basin. The highways converge on it and make it the major import and export center of the region.

Population of the Rio Madeira basin

There are no reliable estimates of human population in the Rio Madeira basin before the Brazilian conquest of the region in the seventeenth and eighteenth centuries. By the last quarter of the nineteenth century, most of the Amerind groups in the Madeira valley were reported by Keller (1874) to have retreated to the right bank tributaries, though the Caripunas were still found to some extent along the rapids in the upper course of the river. Today only a few, small Amerind groups remain in the Rio Madeira basin, and these are located mostly in the headwaters of the affluents. Their future, however, is seriously threatened by the new highway network and the greed of colonizers invading their lands.

The population of the Rio Madeira, subsequent to the decimation of the Amerinds, probably began to grow a little during the rubber boom when latex collectors spred out along the main river and its affluents. Most of the potential rubber collectors that fled Northeastern Brazil, especially the state of Ceará, migrated to the Rio Solimões or Rio Purus where there were more rubber trees and probably less disease. The Rio Madeira had long been known as an unhealthy river, as reported in almost every account that dealt with the region (e.g. Gibbon & Herndon 1854; Keller 1874), and this may have detracted a greater influx of rubber collectors at the end of the nineteenth and beginning of the twentieth century. Nevertheless, almost all of the small settlements along the rivers of the Madeira valley trace their origins to rubber tree sites (called *seringal* in singular or *seringais* in plural). Rubber latex is still the most important plant product exported from the region (IBGE 1979).

Until the opening of the Cuiabá/Porto Velho highway, the Rio Madeira valley was too isolated to spur urbanization. The old railway towns of Porto Velho and Guajará-Mirim, however, would serve as the main urban foci once the road connection was made and large-scale immigration began. In 1950 Porto Velho could boast no more than about 12,000 inhabitants, by 1960 about the double of this, and by 1970 an estimated 50,000 people were living in the frontier city (IBGE 1975b). The 1970's witnessed the explosive growth of Porto Velho as slums sprang up around the outskirts of the center, and by 1975 there were over 100,000 people in the city (COGET 1978). A new census is being carried out in 1980 and is expected to report more than 150,000 Porto Velho residents. Guajará-Mirim is the second largest city in the Madeira valley (in which I am including the Rio Mamoré) and presently has an estimated population of about 25,000, and, like Porto Velho, is still growing very rapidly.

Riparian settlement along the Rio Madeira appears to be sparse compared to the Rio Solimões-Amazonas, and this is probably due to the following factors: a small floodplain and hence limited agricultural potentials; greater isolation from urban markets; and possibly, still a high incidence of malaria which discourages riparian settlement.

Four (Calama, Manicoré, Novo Aripuanã, and Borba) of the six towns and villages located on the banks of the Rio Madeira were not, as of 1980, connected by road with the Amazonian highway network. There were plans, however, to build roads to all of them and thus end their river isolation.

The major focus of colonization in the Rio Madeira basin is in Rondônia along the Cuiabá/Porto Velho highway, and there were an estimated 50,000 families along this frontier axis in 1980, making it even more important as a colonization region than the Transamazon highway. The towns of Ariquemes, Jiparaná, and Vilhena, located along the Cuiabá/Porto Velho highway in Rondônia, all have more than 10,000 inhabitants.

Animal protein flow in the upper Rio Madeira valley

In the late 1970's, fish was the most important animal protein source in most of the western Amazon. Giugliano et al. (1978) studied diets in various neighborhoods in Manaus and reported, as market data also indicated (Petrere 1978), that per capita consumption of fish was very high in the capital city. In lower income groups they estimated per capita consumption of fish to be about 150 grams per day. With recent fish shortages in Manaus, the figure is probably somewhat lower at present, but still relatively high in comparison to other animal protein sources. Smith (1979) estimated that per capita consumption of fish in Itacoatiara, a city of about 30,000 inhabitants on the middle Rio Amazonas, was on the order of 104 grams per day, making it more important than other animal protein sources. The picture outlined above constrasts sharply with the situation in Porto Velho where, since at least 1974, imported beef has been the most important animal eaten in Rondônia's capital.

Porto Velho began importing beef from Mato Grosso after the opening of the Cuiabá/Porto Velho highway. Cattle ranching had expanded greatly in the area around the Pantanal in Mato Grosso in the 1960's and 1970's (IBGE 1977), and thus, with the opening of the highway, Porto Velho had access to an alternative protein source other than the local fisheries that could not meet the necessary demands of the rapidly growing city. Beef cattle imported from Mato Grosso accounted for at least 63 percent of the total weight of animal protein consumed in Porto Velho in 1977 (Goulding 1979). In 1978, the Government of Rondônia relaxed cattle import restrictions, and the region began to buy nearly all of its beef from Bolivia. The Bolivian pastures are located between the Rio Beni and the Rio Pilcomayo, and beef is trucked or shipped by boat to Guajará-Mirim on the Rio Guaporé. From Guajará-Mirim the cattle are trucked to Porto Velho. About 95 percent of the beef consumed in Porto Velho in 1979 originated in Bolivia (data supplied by Fri-Rondon, Porto Velho), and this animal protein source represented no less than 85 percent of the total weight of all the animals consumed in the city in that year.

Local information gained from fishmongers indicated that Porto Velho never had a very reliable fish supply. During the 1950's and 1960's, the market was often glutted during the low water season when fish could be caught easily, but during most of the rest of the months locally caught fish were scarce. During this period buyers were purchasing fish in Manaus and Belém and shipping it upriver to Porto Velho. Fresh fish were also imported in DC-3's that supplied Rio Branco, Acre as well. With the opening of the Porto Velho/Manaus highway in the early 1970's, Porto Velho fishmongers began to truck fish purchased in Manaus, Itacoatiara, and Manacapuru (see Fig. 2.3), especially between February and June when the river is high and the Rio Madeira fisheries are at a minimum. In 1977 and 1978 over one-third of the total fish consumed in Porto Velho was imported from the cities in the state of Amazonas mentioned above. In 1980, the Government of Amazonas prohibited the exportation of fish from the state because of its scarcity in the local markets during the high water season.

During the 1970's there was a chicken for catfish exchange between the southeastern and southern states of Brazil and Rondônia. When large scale catfish operations began in the Rio Madeira in the early 1970's, scaleless fish had a very poor market locally, but refrigeration companies realized that catfish filets would fetch handsome prices in São Paulo and Paraná. In 1977, about 212 tons (cleaned fish) of catfishes were exported to São Paulo and Paraná, but more than double of this amount, in weight, of chickens were brought back in the same trucks. In one sense, then, Porto Velho may have benefited by exporting its catfish, in that truckers brought back more chicken, though admittedly it was sold at a higher price than would have been catfish locally.

Nature of fisheries in relation to urban centers

In the late 1970's, the Rio Madeira basin offered a variety of cutural situations casting light on the evolution of Amazonian commercial fisheries. These ranged from subsistence villages largely dependent on fish for their main source of animal protein, to larger urban entities, such as Porto Velho, that were still heavily exploiting the local fisheries but that had already become dependent on alternative protein sources. I will briefly discuss this sequence here, because I think it will one day be shown to be more or less true for most of the western Amazon.

Small riparian settlements and villages with populations less than about 3,000 people – especially those that are not connected by road to one of the highways – are largely dependent on subsistence fisheries for their local protein. The men, and also many of the women, are highly skilled fishermen. Traditionally they have relied on indigenous fishing methods such as bow-and-arrow, pole-and-line, harpoon, castnet, and gig, but today the gillnet is becoming the dominant gear. These villages do not support a fish market per se, and transactions in fish are more on a personal and kinship basis. Along the Rio Madeira most of these villages are located near the mouths of the larger tributaries or rainforest streams, and this places them – and not all together accidentally – in close proximity to a wide variety of fishing habitats, including flooded forests, beaches, floodplain lakes, and the affluent debouchures.

As the river villages grow in population, and economic activity becomes more diversified and more money is available, commercial fishermen and fishmongers become established. The fisheries are usually located within a short distance of the market as ice is unavailable and catches must be sold within a relatively short period of time to prevent spoilage. At this stage in development of the fisheries, seines become important and large fish schools are attacked. Seine fishing without an accompanying ice boat usually leads to a high spoilage rate as fishermen overfish in relation to the daily demands of the limited market. Fifty percent losses, especially during the low water season, are not uncommon on many days. The city of Humaitá was at this stage in the late 1970's.

The last step, thus far, in the evolution of the Rio Madeira fisheries is the building of a permanent fleet equipped with ice boats that can travel long distances. Porto Velho and Guajará-Mirim had reached this point in 1980. The fisheries are intensively exploited, both for the local and export markets. Seines and gillnets become the most important fishing gear. In the case of the Rio Madeira, which is dominated by the Porto Velho fleet, total catches began to decline after a few years of intensive fishing and thus no further growth in the commercial fisheries is to be expected. The local government, in fact, has already begun to prohibit the exportation of fish during periods when there are critical shortages of animal protein in the Porto Velho market.

CHAPTER 3

The fisheries

Catfish fisheries

The Rio Madeira region has about a dozen species of large catfishes – all of which are discussed individually in Chapter 5 – but traditionally these have not been very important in the commercial fisheries. This is due to a wide variety of other fishes being available, especially the characins and cichlids that are preferred food fishes locally, and to the reputed pathogenic properties of catfish flesh. Brazilian refrigeration companies must be given credit for encouraging the exploitation of the large catfishes in the Rio Madeira valley, beginning in about 1970 on a large scale. When the Cuiabá/Porto Velho highway was opened, refrigeration companies from Southern and Southeastern Brazil looking for investment opportunities in the Amazon, were quick to realize that the large catfishes could be trucked to states such as São Paulo and Paraná where they would fetch at least ten-fold the selling price of Rio Madeira fishermen.

The most important siluroid* fishery in the Rio Madeira basin has been at the Cachoeira do Teotônio, or the second major cataract above Porto Velho. The annual migrations of catfishes through these rapids has long been known (see below), and it was at this location where refrigeration companies first encouraged local fishermen to catch large quantities of siluroids. This fishery will be discussed in detail because of its uniqueness, after which will follow a discussion of how the rest of the river, free of rapids, is exploited for catfishes. In a later section in this chapter, it will be shown how the siluroid fishery frontier has shifted to the Rio Guaporé and Rio Mamoré.

The Teotônio rapids

The Teotônio rapids lie about 20 km upstream of the first of the Rio Madeira cataracts. Archaeological sites and dark, humus rich soils (*terra preta do índio*) near the Teotônio rapids, indicate that the area was settled and farmed by many generations of Amerinds. Auspicious fishing conditions probably encouraged settlement near the cataracts, but the historical documents – which are meager for this region of the Amazon – are silent on the matter of indigenous fishing techniques at the Rio Madeira rapids. Gonçalves da Fonseca, a Portuguese explorer who passed through the cataracts in 1749, appears to be about the first person to make a recorded statement about the fish life at the Madeira rapids. Said Fonseca, 'The fish that one finds after entering the *cachoeiras* is much more delicious than that caught downstream' (cited by Ferreira 1959). He was probably speaking of catfishes, as these are the most abundant fishes at the rapids during most of the year.

Franz Keller (1874), who we saw earlier was employed by the Brazilian government to survey the Rio Madeira cataracts, was impressed by what he

---

*The term siluroid is synonymous with catfish.

saw of fish life at the Teotônio rapids. Interjecting the opinion that he thought the use of fish poison, a habit native to the region, was worthy only of barbarians, he would not have hesitated to use it at the Teotônio rapids. He was disgusted with what he saw at Teotônio. In Keller's words:

'It was at the Salto de Theotonio, the most considerable of the cataracts of the Madeira, where a rugged reef 10 meters height crosses the river-bed. A great number of pools had been left by the receding floods in its holes and on shore, just about where fish probably had tried to pass the fall in lateral channels, or by leaping and bounding over the breaks to continue, in the smooth above, their search for an appropriate place to deposit their spawn. In the largest of these pools many hundreds of gigantic fish had been cut off from the main stream, perhaps weeks before our arrival, and were dying slowly in the warm water of the basin, which was impregnated with every variety of putrid matter. We counted already more than five hundred bodies of large dead fish in every stage of decomposition, floating upon the surface of the slimy green water, and emitting pestiferous exhalations. From time to time a huge *surubim* rose from the depth and moved slowly, almost torpidly, through the thick element. Some dozens of black vultures *(urubus)* looked sharply and anxiously at us, and at the foul pond, their richly laden table, the while sitting rigid and motionless on the neighboring rocks with their wide wings opened to the evening breeze, probably to air their feathers. They reminded us, in their immobility, of bronze eagles on the crown of some old towers. In spite of the sickening aspect, we had the greatest difficulty to prevent our Indians from harpooning the half-dead fish and making themselves seriously ill with this nauseating food, although they had, with but little trouble, succeeded in taking a large quantity of wholesome fish below the fall, in the bays and creeks of the shores, and at the mouth of the small rivulet' (1874: 85-86).

According to local residents whose families trace their roots at the Teotônio cataract back to the 1910's and 1920's, the rapids were first fished commercially in the 1930's. In that decade a few tons of salted catfish were sold to river traders servicing rubber collectors. Residents state that 1948 witnessed the first motorized canoe negotiating the Santo Antonio rapids near Porto Velho and reaching Teotônio, and in subsequent years, especially during the low water season, fresh fish (mostly characins) was transported to the small market in Porto Velho. Catfishes continued to be caught and salted for rubber collectors until the mid-1960's when tin-stone prospecting and mining began in earnest and fishermen and rubber collectors were lured from their normal professions to the cassiterite fields in search of quick wealth. When the Federal Government prohibited individual prospecting in 1971, the independent miners were forced to abandon their quarries.

Some of the ex-fishermen gravitated to the Teotônio cataract where refrigeration companies offered to buy catfishes, and they were joined by a few highly skilled fishermen who had fled São Paulo and Mato Grosso when those state governments began to clamp down and control their fisheries. It is perhaps an ironic note that the most important catfish species (*Brachyplatystoma flavicans*, Pimelodidae) is known vernacularly as the *dourada*, or the gilded one, and indeed the Teotônio fishermen became known in the area as 'piscine miners'. As this story will reveal, however, the whiskered gold, within a few years, would begin to get rarer in the Rio Madeira, and the fishermen-miners would begin to move on to new piscine and mineral frontiers.

Of all the major fish groups exploited commercially in the Rio Madeira valley, catfishes are the only one that make large-scale upstream migrations during both the high and low water seasons (Table 3.1). If not for the Teotônio rapids, however, catfishes would be little vulnerable to exploitation during the floods because of the large amount of wood transported by the Rio Madeira during its annual inundation and its greater depth and expanded width at this time of the year. At the Teotônio cataract commercial fishing goes on all year, with September through March being the most productive period; this embraces the period between the lowest water point and the height of the flood. Within this period, however, there appears to be a bimodal intensity of upstream catfish migrations, with the first centering on the low water months, September

Table 3.1 The catfishes exploited at the Teotônio cataract and the months that each species migrates upstream and through the rapids.

Fish species	Schools pass through cataracts
	J F M A M J J A S O N D
Pimelodidae	
Dourada (*Brachyplatystoma flavicans*)	▬▬▬▬▬▬ ▬▬▬▬▬▬
Piraíba/Filhote (*Brachyplatystoma filamentosum*)	▬▬▬▬▬▬
Babão (*Goslinia platynema*)	▬▬▬▬▬
Caparari (*Pseudoplatystoma tigrinum*)	▬▬▬▬▬▬
Surubim (*Pseudoplatystoma fasciatum*)	▬▬▬▬▬▬
Jaú (*Paulicea lutkeni*)	▬▬▬▬▬▬
Pirarara (*Phractocephalus hemiliopterus*)	▬▬▬▬▬▬
Peixe Lenha (*Surubimichthys planiceps*)	▬▬▬▬▬
Dourada Fita (*Merodontotus tigrinus*)	▬▬▬▬▬
Pintadinho (*Callophysus macropterus*)	▬▬▬▬
Barba Chata (*Pinirampus pirinampu*)	▬▬▬▬
Piramutaba (*Brachyplatystoma vailantii*)	▬▬▬▬▬
Bico de Pato (*Sorubim lima*)	▬▬▬▬
Mandi (*Pimilodus* spp.)	▬▬▬
Doradidae	
Bacu (*Megaladoras irwini*)	▬▬
Bacu Comum (*Pterodoras granulosus*)	▬▬ ▬▬
Bacu Pedra (*Lithodoras dorsalis*)	▬▬
Cuiu-Cuiu (*Oxydoras niger*)	▬▬

through November, and the second from the end of December to the end of April.

By far the most important commercial catfish is the *dourada* (*Brachyplatystoma flavicans*, Pimelodidae), which in 1977 represented about 67 percent of the total catch at the Teotônio rapids, and the contribution of the species appears to have been on a similar order since then. The *dourada* moves upstream during both the high and low water seasons (the reasons for these migrations will be discussed in Chapter 5). Of less importance is the *piraíba*, or *filhote* for smaller individuals of the same species (*Brachyplatystoma filamentosum*, Pimelodidae), which moves upstream mostly during the low water period, and the *babão* (*Goslinia platynema*, Pimelodidae) that migrates during the floods. Fishermen's reports and records of refrigeration companies indicate that fishes of the genus *Pseudoplatystoma* were important in the initial years of catfish exploitation at the Teotônio cataract, but that yields have declined considerably since about 1976. The other species are caught in much lesser quantities, though there are years when large schools of *piramutaba* (*Brachyplatystoma vaillantii*, Pimelodidae) appear at the cataract and are heavily exploited. This was the case in 1974 and 1979.

The Teotônio rapids are about 300 m in length, and the river is about 700 m across at the crest of the cataract. As mentioned, a rocky bulwark shoots across the width of the river, about three-fourths of which is emerged during the lowest water period. At intermediate water levels, the Rio Madeira is split into several currents that rush violently through the lower parts of the Teotônio bulwark, while during the floods the onslaught of water nearly covers the entire rocky fundament. The upstream moving catfishes attempt to find the routes of least resistance where they can negotiate the violent currents. During the floods they must seek the banks where frictional drag slows the current, and they do not appear to be able to pass through the main part of the cataract. At lower water levels there are often two or three currents which allow the fish passage upstream, but there are others that form mini-falls which allow no further entrance. Teotônio fishermen know the topography of the cataract so well that every current and architectural feature of the rocky bulwark has been given a name. With these place names, the Teotônio fishermen are able to predict, and quite successfully I might add, the approximate date and exact location where fish schools will appear in an attempt to pass the rapids. Armed with this knowledge, based in large part on oral history from long-time residents and about a decade of commercial fishing, the fishermen have adopted a series of fishing methods to confront the exigencies of the violent water they fish. Each of these methods will now be discussed.

Gaff

The gaff, or *fisga*, is the most important device used to catch catfishes near the crest of the Teotônio cataract during the high water season. The catfish gaff is constructed from a 5-8 m long bamboo pole, a large barbed hook with about a 15 cm shank, a pulling cord, and a string which is used to fasten the hook to the pole (Fig. 3.1). The gaff-hook does not have an eye, but the proximal part of the shaft is flattened with a small rounded head at the end. To this part of the hook is tied an attaching line which consists of a series of overlapping half-hitches, and each of these knots communicates with a loop a little larger than the hook. A short piece of line is left dangling from the first half-hitch, and this is used to fasten the hook to the bottom side of the end of the bamboo rod. The loops in turn are gathered together and one end of the pulling cord is tied around them; the pulling cord is then wrapped two or three times around the pole, with most of its length being left free of the gaff-rod.

During the low water period, Teotônio fishermen build platforms along the left bank of the cataract (Fig. 3.2). These fishing platforms are wedged into the crevices of outcropping rocks and securely fastened to the shore with ropes. The platforms allow the fishermen to perch themselves several meters out into the current during the floods when the upstream moving catfishes are forced to the left bank to gain passage through the violent rapids. When schools of catfishes are climbing the left bank current, fishermen stroke the water with their gaffs until sinking the hook into a siluroid victim. Once the catfish is gaffed, its weight and fight in the strong current serve to pull the hook off the end of the bamboo pole. The fisherman then frees his pole and pulls the fish in with the pulling line. If the fish is too big to be manhandled and the fisherman cannot maintain his grip, the pulling line is freed, but the fish usually does not escape because one end of the cord is tied to the platform or a rock. Once the fish has tired, the fisherman then pulls it in. In January, February, and March gaffing often goes on day and night while the big catfishes are moving upstream. *Dourada* (*Brachyplatystoma flavicans*, Pimelodidae) account for most of the catfish that are gaffed at Teotônio, and this species begins moving through the left bank current at the end of December, and these migrations last until about the end of February. In February schools of the *babão* (*Goslinia platynema*, Pimelodidae) appear at the cataract and use the same left bank route as the *dourada*, and they are gaffed until about the end of April when their migration through Teotônio terminates. By May there are few catfishes moving through the rapids, and river level has dropped to the point where several currents can be negotiated other than the left bank one, and the gaff is no longer effective.

The upstream migrating catfishes that move through the Teotônio rapids attract not only commercial fishermen but also voracious smaller catfishes that attack their larger relatives when they become concentrated below the crest of the rapids. These predators are a nuisance to fishermen because they often destroy gaffed fishes before they can be landed. A potential catch can literally be eaten by schools of these fishes in less than a minute, and the fisherman is rewarded with no more than the head and the skeleton of the large catfish he has gaffed. I have identified five genera, in three families, of catfishes that are involved in attacks on large siluroids at the gaff fisheries of Teotônio. The smallest predators belong to the genera *Pseudostegophilus* (Fig 3.3) and *Pareiodon* of the family Trichomycteridae. These fishes have inferior mouths endowed with teeth that allow them to rip or rasp out pieces of flesh. *Cetopsis* (Fig. 3.4) and *Hemicetopsis*, popularly called the whale catfishes (family Cetopsidae) in English because of their gross resemblance to the

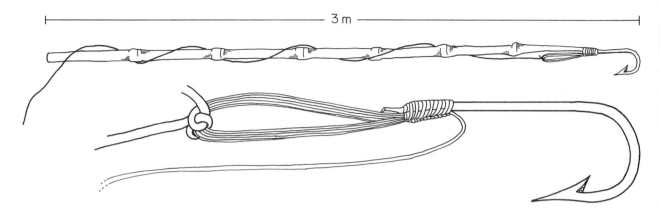

Fig. 3.1 The gaff used at the Teotônio rapids to catch large catfishes.

Fig. 3.2 Gaffing site at the Teotônio cataract of the Rio Madeira. The gaffed fish is the *dourada (Brachyplatystoma flavicans*, Pimelodidae).

Fig. 3.3 Candirú pintado *(Pseudostegophilus* sp., Trichomycteridae) that attack large catfishes at the Teotônio cararact. Above: Profile. Below: Ventral view showing position of mouth of the *candirú pintado*. About 12 cm fork length.

Fig. 3.4 Candirú-açú *(Cetopsis* sp., Cetopsidae), a voracious predator that attacks large catfishes at the Teotônio rapids.

giant cetaceans, range from about 15-25 cm when adult. Their rounded, smooth bodies allow them to insinuate themselves into ripped and torn flesh, all the time biting away at the fish in which they are wedged. It is not uncommon for a fisherman to pull in his catch and find several of these voracious cetopsids inside the body cavity of the large catfish, all of course, still alive, and still attempting to feed out of the water. The last identified culprit in these attacks is a somewhat larger catfish, *Callophysus macropterus* (see Fig. 5.17) of the family Pimedodidae. This bewhiskered and spotted catfish is armed with incisive-like teeth which also allow it to rip out pieces of a victim's body. It is still unclear whether these predators attack the upstream moving catfishes in such a voracious manner when they are not being gaffed. Under more natural conditions, they may lead a more scavenger-like life, cleaning up on aquatic carrion.

Handline

Handlines, consisting of but a line and a baited hook, are commonly used in subsistence fishing along the riverbanks to catch the smaller catfishes. They are not very effective, however, in capturing the larger catfishes in the deep channel as too much time is required for a large fish to find the bait, and trotlines – which can be left unattended for many hours – are more effective because a larger number of hooks can be employed. At the Teotônio rapids, however, the handline is effective in catching the large *jaú* (*Paulicea lutkeni*, Pimelodidae). This blubbery catfish appears at the rapids apparently to feed on the medium sized characins (20-40 cm average) that become concentrated below the crest of the cataract during the low water season. More than any of the other large catfishes, it appears to actively pursue prey in the turbulent waters; most of the other large catfish species do not feed below the rapids but try to find a route to move upstream. When fishermen see the *jaú* feeding in the rapids, they throw the baited hook into the same general area. Once the large fish takes the bait and is hooked, it is then pulled onto the rocks and stunned with a heavy club (Fig. 3.5). Occasionally the *piraíba* (*Brachyplatystoma filamentosum*, Pimelodidae) is also taken with the handline, but most of the other species only rarely so.

Wire traps

Fiber traps shaped like cones are widely used in the Amazon basin to catch turtles, and the Teotônio catfish trap is similarly constructed but much larger and uses strong wire instead of plant materials (Fig. 3.6). The opening of the catfish trap has about a 2 m diameter, while the length of the main body is about 2-3 m. The opening of the *côvo*, as the trap is called, is placed facing downstream in the current where catfishes are moving upstream along the shore. Upstream migrating fishes enter the mouth of the trap and then pass through a narrow opening that is armed with backward directed, stiffened pieces of wire pointing slightly inward on each other. Once a fish enters the main chamber of the trap, it is unable to turn around or back out. A door at the back of the trap allows the catch to be removed. The *côvo* is effective for catching fishes less than about one meter in length.

Castnet

Castnets are effective during the low water period when rocky outcrops of the cataract form islands and allow fishermen to perch themselves over pools and currents where upstream moving fishes become concentrated (Figs. 3.7 and 3.8). The Teotônio castnet has about a 6-8 m diameter when opened and carries up to 15 kg of lead around its circumference (Fig. 3.9). Mesh sizes vary between about 10-18 cm, and the handcord is about 20-30 m in length. The handcord of the cataract castnet is usually not tied around the wrist, as is the custom when using smaller models in subsistence fishing. Instead the long handcord is either fastened to a shore support or more often strung out behind the fishermen. With the handcord free, the fisherman does not risk being pulled into the water by a large fish that gets entangled in his castnet.

In preparation for the cast, the upper two-thirds of the net is folded in double or triple to allow for a full swing. Then a piece of the circumference of the net is held in the teeth, or in the case of the not infrequent edentulous fisherman, between the neck and head, while the free hand, palm down, gathers about half of the weighted line from the inside. The net is now in postition to cast and is swung backwards to give it momentum for its forward projec-

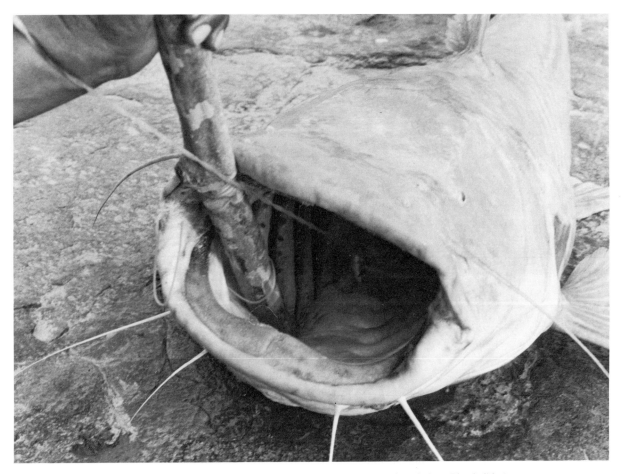

Fig. 3.5 A fisherman prepares to remove his handline hook from the stunned *jaú* (*Paulicea lutkeni*, Pimelodidae).

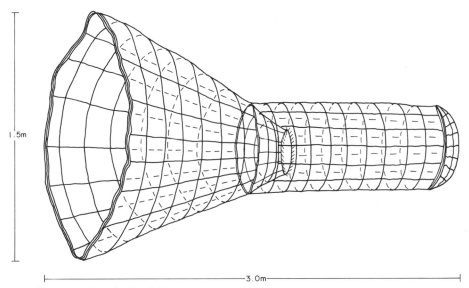

Fig. 3.6 The *côvo*, or fish trap, used at the Teotônio cataract.

Fig. 3.7 The Ceará rock in the middle of the Teotónio cataract. Fishermen perch themselves around the edge of the rock platform to throw their castnets on top of upstream moving catfishes.

Fig. 3.8 Close-up view of Ceará rock (see Fig. 3.7) showing fishermen with large catfishes that have been caught in the turbulent water.

Fig. 3.9 Fisherman pulling in a *jaú* (*Paulicea lutkeni*, Pimelodidae) with a castnet.

tory. As the castnet is swung forward the fisherman releases his three grips and the bag opens in mid-air and falls on the water with its greatest area extended. The heavy weights carry the net to the bottom, and, as the handcord is pulled by the fisherman, the weighted line comes together and encloses the fishes that are within. The large mesh usually entangles the catfishes behind the gills or by the paired and dorsal fins. The catch is then pulled onto dry ground and clubbed.

The River Channel

As of 1980 the entire Rio Madeira, from the Teotônio cataract to the Rio Amazonas, was being exploited for large catfishes. Porto Velho fishermen, however, only go about as far as the Rio Aripuanã, whereas the lower course of the river is exploited mostly by fishermen from Manaus and Itacoatiara. In the main river channel the large catfishes can only be caught in large quantities near the bottom, and there are only two effective types of gear for capturing them in this habitat.

Drifting deepwater-gillnet
The *caçoeira*, or drifting deepwater-gillnet, was first introduced into the Rio Madeira basin in about 1970, and since then its use has proliferated to an extent to make it the most important method for catching large catfishes (Figs. 3.10 and 3.11). The *caçoeira* ranges in length from about 100-300 m and in height from about 3-8 m; mesh sizes vary between 14-25 cm. Either transparent monofilament or nylon line is used for construction. The bottom of the net is heavily weighted so that it will sink to the bottom of the river, often to 20 m or more, while to the top horizontal line are strung enough floats to maintain the net vertically without floating it to the surface. Thick ropes are tied to the top corners, with one of these being fastened to a large block of styrofoam or other floating material such as empty plastic gas cans, while the other rope is manipulated by a fisherman in a canoe. The use of a large float eliminates the need of a second canoe, and two fishermen, one paddling and one paying the net and manipulating it with the pull line once set in the water, can operate the drifting deepwater-gillnet efficiently. Usually both fishermen, however, pull the net in to check it for fish.

The drifting deepwater-gillnet can be used at any time of the year in the Rio Madeira, though its effectiveness varies greatly according to water level, location, and fish migrations. It is most productive during the low water season when river level is 7-13 m below the height of the flood. At this time there is much less wood being carried downstream – the main hinderance because of snags – and fishes are more concentrated in the restricted channel. The beaches are also well defined at this time, and they are excellent habitats across which to drift the *caçoeiras*, especially at night when the catfishes move more closely to shore in search of prey. During the low water period, when characins are migrating upstream (Chapter 5), the large catfishes pursue them as prey. When *Caçoeira* fishermen spot a school of upstream migrating characins, they often pay their nets out a few hundred meters upstream of the school. Because the characins are smaller than the mesh of the drifting deepwater-gillnets, they pass through, but the larger catfishes following behind their prey, are caught.

When characins migrate down the tributaries to spawn in the turbid water of the Rio Madeira, the large catfishes become concentrated near the confluences of the affluents and the principal river where they feed heavily on the migrating ripe fishes. Fishermen are aware of this behavior, and likewise, concentrate their efforts near the confluences of the tributaries.

In the river channel between the Teotônio cataract and Porto Velho, a distance of about 20 km, the large catfishes appear to become concentrated. This appears to be due to the cataracts, but it is not understood why the catfishes delay their journey when arriving near the rapids. The drifting deepwater-gillnet is very effective in this stretch, and in 1977 accounted for over 50 percent of the total catfish catch of the cataract area between Porto Velho and Teotônio.

River channel trotline
The main line of the *grozeira*, or river channel trotline, consists of 100-300 m of heavy nylon or monofilament cord, to the last one-third of which is at-

Fig. 3.10 The drifting deepwater-gillnet in the Rio Madeira channel. See below for underwater view.

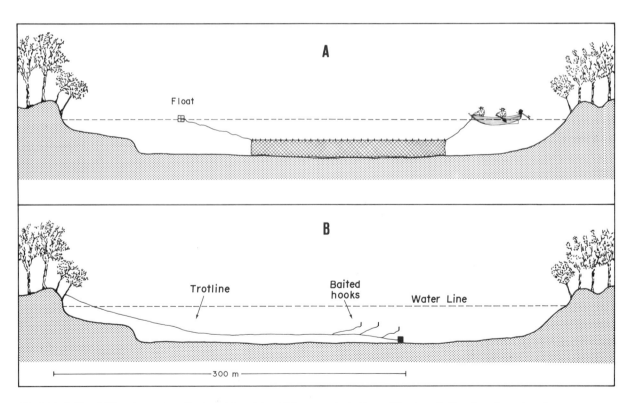

Fig. 3.11 A. The drifting deepwater-gillnet used to catch catfishes near the bottom of the river. B. The river channel trotline.

tached a series of 1-3 m lines, each of which has a 8-15 cm shanked hook (Fig. 3.11). One end of the *grozeira* is attached to a riverside shrub or to two or three poles driven into the mud, thus when a large fish is hooked it is 'played' by the flexibility of these shore holds in a manner similar to a fishing pole. To the other end of the line is tied a heavy weight, usually a rock. The hooks are most often baited with characins, those of the genera *Prochilodus*, *Semaprochilodus*, *Curimata*, *Triportheus*, and *Brycon* being the favorite. To set the *grozeira* in position for fishing, the line is strung out perpendicularly from the shore in a canoe, and the rock is dumped overboard in the deep channel and the baited hooks sink to the bottom. After being hooked, the large catfish carries the rock about until becoming exhausted, but it seldom dies in this futile struggle. The fisherman begins at the shore, slowly pulling the line up until reaching the hook-lines; at this point the hooked catfish usually makes its last attempt at freedom and the fisherman must be ready for the final lunge lest he be pulled into the river and perhaps hooked himself. Once the catfish is pulled to the surface, its head is beaten with a heavy club until the catch is stunned and can be brought aboard. I have witnessed, in total admiration, fishermen who check their trotlines while seated in a small canoe with no more than 5 cm freeboard. Their skill enables them to do battle with 100 kg fishes and, more remarkable, considering the small size of their canoes, they are able to wrestle the large fish into their craft (Fig. 3.12).

Trotlines are the only effective devices for catching larger fishes, of say, over 50 km. The drifting deepwater-gillnets have too small of meshes to entangle the largest individuals. Not all of the catfish species are highly susceptible to trotlines. The *piraíba (Brachyplatystoma filamentosum*, Pimelodidae) is by far the most common species captured, follwed by *pirarara (Phractocephalus hemiliopterus)*, and *jaú (Paulicea lutkeni)*. The *dourada (Brachyplatystoma flavicans*, Pimelodidae) is only rarely caught, even though it is the most abundant of the large catfishes of the Rio Madeira.

There are two problems which greatly reduce the effectiveness of trotlines, namely, that considerable time must be spent catching bait, and small predaceous catfishes constantly remove the bait before a large fish has a chance to find it.

Migratory characin fisheries

All of the important food fish characins of the Rio Madeira basin – and also of the Central Amazon centering on the Rio Solimões-Amazonas – are migratory fishes in the sense of forming large schools and migrating in the rivers at some time of the year (Table 3.2). These migrations are well known to commercial fishermen, and major seasonal fisheries are based on them. Fishes have never been tagged successfully in the Amazon, and most of our information about migrations is based on data from the commercial fisheries. This data, in the case of the Rio Madeira, strongly suggests that there are two distinct migrations annually for most of the migratory species. It is from this perspective, then, that the exploitation of the migratory characins will be discussed.

For a point of departure the spawning migrations may be considered first, which begin in about mid-November when river level is rising and last to at least the beginning of February, a month or so before the peak of the annual flood. In November – following the lowest water season – most of the migratory adult characin biomass in the Rio Madeira basin is found in the large rightbank tributaries, but a small percentage is also residing at this time in the floodplain lakes and main channel of the principal river. It is still unclear why these characins prefer the clearwater tributaries vis-à-vis the turbid Rio Madeira in the pre-spawning period; there is at least one species *(Colossoma macropomum*, Characidae), however, that prefers to remain in the Rio Madeira until spawning, but it is aberrant in its behavior. When the spawning time arrives, the migratory characins form large schools and begin to move down the clearwater tributaries to seek the turbid water of the Rio Madeira. The Rio Madeira and its large right bank tributaries rise together with the floods, and thus the principal river does not invade its affluents; consequently the meeting of the two distinct water types is well defined between the mouth of the affluent and the main river. Rio Madeira floodplain areas that are lower than the main river, however,

Table 3.2 Migratory characins of the Rio Madeira.

Characidae
   Jatuarana (*Brycon* sp.)
   Pacu toba (*Mylossoma duriventris*)
   Pacu encarnado (*Mylossoma albiscopus*)
   Pacu branco (*Mylossoma aureus*)
   Sardinha chata (*Triportheus angulatus*)
   Sardinha comprida (*Triportheus elongatus*)
   Tambaqui (*Colossoma macropomum*)
   Pirapitinga (*Colossoma bidens*)

Prochilondontidae
   Jaraqui escama grossa (*Semaprochilodus theraponura*)
   Jaraqui escama fina (*Semaprochilodus taeniurus*)
   Curimatá (*Prochilodus nigricans*)

Curimatidae
   Branquinha chora (*Curimata latior*)
   Branquinha cabeqa lisa (*Curimata altamazonica*)
   Branquinha comum (*Curimata vittata*)
   Cascudinha (*Curimata amazonica*)

Anostomidae
   Aracu cabeqa gorda (*Leporinus friderici*)
   Aracu botafogo (*Schizodon fasciatus*)
   Pião (*Rhytiodus argenteofuscus* and *microlepis*)

Hemiodontidae
   Orana (*Hemiodus* spp.)
   Flecheiro (*Anodus elongatus*)

suffer an invasion of turbid water at the beginning of the floods, and it appears that at least some of the migratory characins may spawn in these low transparency waters on the floodplain instead of moving out into the flowing channel. Nearly all commercial fishing effort in the Rio Madeira basin at this time of the year is concentrated at the mouths of the large rightbank tributaries.

There is a sequence to the spawning migrations out of the rightbank tributaries that I observed during three different years, and fisheries data also support this. The first species to descend the affluents to spawn in the turbid waters of the Rio Madeira was always the microphagous feeding *jaraqui escama fina (Semaprochilodus taeniurus*, Prochilodontidae), and in all three years (1977-1979) it appeared at the affluent mouths in mid-November. Following this species came *jaraqui escama grossa (Semaprochilodus theraponura)*, and these spawning migrations lasted until mid-December. Next descended *curi-matá (Prochilodus nigricans*, Prochilodontidae), and *branquinhas (Curimata* spp., Curimatidae), whose migrations lasted until mid-January. It should be noted that the first characin species to descend to spawn are microphagous feeders, meaning fishes that remove fine particles from the bottom or other substrates such as submerged trees and limbs. By the third week of December many characin species are descending, the most important in the commercial fisheries being *jatuarana (Brycon* sp., Characidae), *pacu toba* and *pacu vermelho (Mylossoma duriventris* and *Mylossoma albiscopus*, Characidae), *sardinha chata* and *sardinha comprida (Triportheus angulatus* and *Triportheus elongatus*, Characidae), and *aracu botafogo (Schizodon fasciatus*, Anostomidae). The large *pirapitinga (Colossoma bidens*, Characidae) was the last species to be seen descending, and this was in the last week of January and first week of February.

Most of the commercial catch of spawning migratory characins in the Rio Madeira basin is taken in about a ten day to two week period – at least in 1977 and 1978 when catches were carefully recorded – when water level is rising rapidly from about eight to four meters below the height of the flood. This accelerated rise in river level presages the floods without apparent danger of a subsequent delay of rains.

The most important and sought after species of migratory spawning characin captured near the tributary mouths is the *jatuarana (Brycon* sp., Characidae). This species, which is somewhat troutlike in appearance (see Fig. 5.24), commands and attractive market price and is also easily observed when it is descending the tributaries to spawn in the Rio Madeira. When descending, the *jatuarana* schools make spiraling movements which are called *rodada* by fishermen. The movements cause the surface water to ripple in expanding spiral movements. From having observed the *rodada* on many occasions, my impression is that individual fishes spiral up-and-down in cadense with others in the water column, but this observation must be taken with a grain of salt until the movements can somehow be seen in the water.

During the spawning period, commercial fishermen anchor their boats in the tributary mouths and

Fig. 3.12 The 110 kg *piraíba* (*Brachyplatystoma filamentosum*, Pimelodidae) was captured with a trotline. The fisherman wrestled the large animal into his canoe while in the middle of the Rio Madeira channel.

Fig. 3.13 The *rede de lance*, or seine, used in Rio Madeira fisheries.

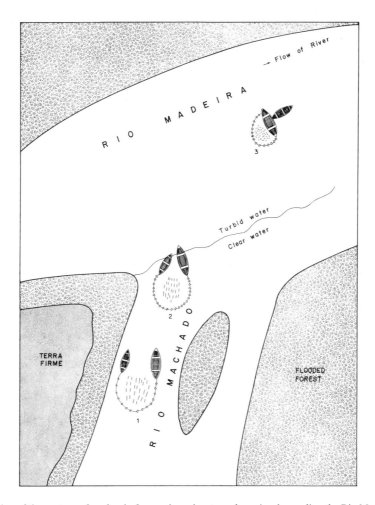

Fig. 3.14 Schematic drawing of the capture of a school of spawning migratory characins descending the Rio Machado to breed in the Rio Madeira. 1) The seine is first payed upstream of the downstream moving fishes. 2) The seine is closed before the school reaches the turbid water of the rio Madeira. 3) The strong current of the Rio Madeira carries the fishermen and their catch downstream; a large fishing boat is then dispatched to the capture site and the fishes are brought aboard.

keep a sharp vigilance for downstream moving schools of fishes. Groups of dolphins *(Inia geoffrensis*, Platanistidae and *Sotalia fluviatilis*, Delphinidae) usually betray the presence of downstream moving schools of fishes before the spiraling movements can be spotted. When many dolphins are seen jumping and chasing fishes, it is a good bet that they are close to a large school of migratory characins on which they are feeding. The fishermen then man their canoes and move upstream to position themselves above the downstream moving school. If they frighten the school too soon, it will often flee to nearby flooded forest where it cannot be pursued with seines. Before discussing the actual capture of a school of migratory characins, the seine itself should be described.

The *rede de lance*, or seine, used in many Amazon commercial fisheries is about 100-200 m in length and 7-10 m in height (Fig. 3.13). Mesh sizes vary from a few to as much as 24 cm when only large fishes such as *tambaqui (Colossoma macropomum*, Characidae) are being pursued. The top of the net is laced with a strong rope on which are also strung enough floats to buoy the seine in the water. The bottom, or weighted, line of the seine carries a sufficient quantity of lead to make sure that the net sinks quickly in the water. The ends of the seine are tapered and attached to holding bars that are constructed of wood. To these pieces of wood are attached the pulling ropes. For paying the seine fishermen utilize one large canoe which carries the net, the net-handler, a paddler astern, and another in the prow who calls the shots. A smaller, second canoe usually contains two men, one of whom in the prow paddles and the other, standing in the middle, operates one of the pulling ropes. When fishing in the affluent mouths for spawning characins, fishermen must circle the schools before they get too close to the Rio Madeira for, once in the turbid water, the fishes either dive deep or disperse as they are no longer seen. Once upstream of the descending school, and in position, the fishermen pay out the seine so that it describes a semi-circle (Fig. 3.14). The two canoes are then paddled to a position slightly downstream of the fish school, with communication with the net maintained by long pulling ropes. At the right moment, the water surface is beaten with the pulling ropes and paddles are pounded against the sids of the canoes to frighten the fishes into reversing their direction of movement and hence fall prey to the sack of the seine. The fishermen then quickly pull up the bottom line from both ends, and the seine forms a closed bag and the fishes are unable to escape. Further uptake of the top and bottom lines diminishes the size of the sack, and the catch can eventually be removed with dipnets.

From about the end of January to mid-March, no large schools of characins are seen in the Rio Madeira or its tributaries. At this time of the year, which corresponds to the main floods, most of the migratory characins have already spawned and are found dispersed in the flooded forests where most of their food is procured (Goulding 1980). Just as the microphagous feeding fishes of the genus *Semaprochilodus* are the first characins to spawn with the coming of the floods, they are also the first taxa to move out of the flooded forests and form schools when water level begins dropping in March or April. At this time the rivers are still high – but falling – and the schools of *Semaprochilodus* move down the tributaries in a manner similar to when they were spawning. However, once entering the Rio Madeira, they move upstream in it until choosing another tributary. These schools of *Semaprochilodus* do not appear to remain in the Rio Madeira for more than two or three weeks at this time of year. Fishermen attempt to catch them in the tributary mouths in a manner described above for the spawning fishes. If they are unsuccessful, however, they may follow and exploit the schools until they enter another affluent. Some of these schooll st be very large, and probably have at least 200,000 fishes.

In May, June, and July, the other migratory characins begin descending the tributaries and entering the Rio Madeira to move upstream in it. During these months fishermen have most success exploiting schools in the tributary mouths where they wait for the migrating characins. The migratory characins captured between March and July are referred to as *peixe gordo*, or fat fish, a reference to the large fat deposits that they have built up while feeding in the flooded forests.

As water level continues to drop in August, September, and October, the main upstream migrations in the Rio Madeira commence, and schools of almost all of the migratory species can be seen in the principal river at this time. Collectively these migrations are referred to as the *piracema*, a term which is taken from *língua geral* and means 'meeting of fishes' (Roque 1968) or 'upstream moving fishes' (Verissimo 1895). The largest catches of the year are made during the *piracema* migrations as the river channel is shallow and restricted and fishes can more easily be spotted. Both the open water seine described above and beach seines of similar construction but much longer (300-400 m) are used to catch the *piracema* schools. Beaches are the ideal habitat for seine fishing in rivers as the upstream moving fishes can be circled with the net touching the bottom. Usually eight to twelve fishermen are required to pull a large beach seine into shore with the captured school of fishes. Because the beach area is shallow, the fishes cannot escape by diving underneath the net before the bottom line can be pulled up to form a sack in the seine. It is not uncommon for fishermen to catch ten tons of fish in a single beach seining in the Rio Madeira.

By mid-October, when river level is rising, the *piracema* schools become scarcer in the Rio Madeira. The only species that is regularly encountered in large schools in late October and early November is the *tambaqui (Colossoma macropomum*, Characidae), the largest characin in the Amazon. Unlike the other migratory characins, this species remains in the Rio Madeira until spawning in November or December; in other words, it does not migrate back to the tributaries before spawning. At this time of year the *tambaqui* is the object of intensive fishing effort because it is the only species encountered in large schools and, as well, because it is considered a first class food fish and hence commands a good market price.

Floodplain fisheries

The size of floodplains is the main factor that determines *total* fish productivity in most tropical river systems (Welcomme 1979). This is true for both systems relatively poor in nutrients and those better endowed. The most productive habitats in Amazon river systems are found on the floodplains of turbid water rivers such as the Solimões-Amazonas and Purus where food chains can be built up from at least three major sources of energy, namely, phytoplankton, aquatic herbaceous plants (macrophytes), and flooded forests. There are still no quantitative studies to indicate the relative importance of the three major energy sources for the fish food chain in turbid river systems, but the phytoplankton-zooplankton link is undoubtedly important for young fish, as may well be periphyton and perizoon, or the microfloras and microfaunas attached to aquatic herbaceous vegetation. The nutrient poor river systems, or those often referred to as blackwaters and clearwaters, have very poor plankton production and the development of macrophtes is usually limited in comparison to turbid river systems. The evidence indicates that the flooded forests are the main source of energy for adult fishes in these systems, but young fish are yet to be studied (Goulding 1980). In any case, there are no reported floodplain areas of clearwater or blackwater river systems that support significant annual fisheries. In short, the only floodplain areas that can be intensively exploited on a large scale from year to year are found along the turbid water rivers were an annual injection of nutrients brought out of the Andes support the food chain that leads to the commercial catches.

For the most part, the Rio Madeira has a narrow and high floodplain, and there is only one area above the mouth of the Rio Aripuanã that supports annual fisheries of any significant scale. This floodplain area is called *Cuniã* and is located about 40 km downstream of Porto Velho on the left bank (see Fig. 1.8). The Cuniã floodplain appears to have been formed in a low-lying area across which the main channel of the Rio Madeira wandered during its historical development. The Cuniã floodplain embraces about 10 km^2, most of which is flooded forest with numerous interspersed lagoons and lakes. The largest lake, or Lago de Cuniã, is filled with blackwater draining the nearby *terra firme*, and an outlet from it meanders about 40 km through the rainforest before entering the Rio

Madeira. At the beginning of the floods the Rio Madeira invades this outlet stream and turbid water reaches as far as the main lake and spreads out over most of the floodplain. There is rapid macrophyte development and plankton blooms at this time. As the annual flood recedes, the inundation forests are drained, and the Cuniã *várzea** waterbodies shrink into small lakes and lagoons, many of which are joined to each other by stream-like connections.

The Cuniã *várzea* has been fished commercially for the Porto Velho market since 1958 when the first ice boat entered the floodplain area to buy fresh fish from the local inhabitants. Prior to about 1960, Cuniã also apparently supported a small *pirarucu (Arapaima gigas*, Osteoglossidae) fishery, in which the product was salted for sale to river traders. Overfishing led to the decline of the *pirarucu* fishery, and the species has not recovered since to be exploited on a significant scale. The fishermen of the Cuniã floodplain are the local residents who are also subsistence farmers who spend much of their time, when not fishing, planting manioc and processing it into flour (*farinha*). The Cuniã residents do not allow Porto Velho fishermen to enter their waters and, to prove their territoriality, have several times turned them away at gunpoint.

The Cuniã floodplain fisheries are highly seasonal, with most of the total catch being taken during the low water period when fishes become concentrated in restricted lagoons and lakes. During the floods the fishes move into the flooded forests and are more difficult to capture. I recorded about 25 species that are regularly caught in the Cuniã floodplain, but most of these are in very small quantities (Table 3.3). The most important species is the *tucunaré (Cichla ocellaris*, Cichlidae), and local fishermen told me that it has been the most important fish since commercial operations began in 1958. The *tucunaré* is also one of the most important species captured in the floodplain lakes of the Solimões-Amazonas (Petrere 1978; Smith 1979). Its importance in the commercial fisheries is due both to its success as a piscivore in floodplain systems, its ability to respond (reproductively) to its predation by the fisheries, and its classification as a first class food fish. Other floodplain species are

*Várzea is a regional term synonymous with floodplain.

undoubtedly as abundant, but they are more difficult to capture, as are the *piranhas*, or command too low of market prices, as do the detritivorous curimatids, to be exploited on a large scale.

The techniques of Rio Madeira floodplain fishing are based on a combination of indigenous methods and *caboclo*, or miscegenated Amazonian, inventions developed in the last two or three centuries. These methods rely heavily on the skill of the individual fisherman and his intimate knowledge of the behavior of the fishes. The most important modern introduction has been the gillnet, which has been especially effective for capturing some of the larger fishes. Each of the methods employed in the Rio Madeira floodplain fisheries will now be discussed in detail.

Gig

The *zagáia*, or gig, is a pronged spear usually 1.5-2.0 m in length (Fig. 3.15). The point is constructed from heavy wire or tempered steel, and can have two or three barbed prongs. The gig is an extremely effective fishing device when used at night with a flashlight or other focused light source, as several fish species, especially the cichlids, are nocturnally inactive and remain near the surface where they can be seen. In the Cuniã floodplain, most of the *tucunaré (Cichla ocellaris)* and *cará-açu (Astronotus ocellatus)* catch is taken with the gig. These fishes are most vulnerable to gigging during the low water season when they become concentrated near the shore areas, especially in the narrow band of flooded forest that surrounds the small lakes and lagoons just before and after the minimum water level. The gig fisherman paddles slowly along the edge of the open waterbodies or through the flooded fores, focusing his flashlight into the water. When a fish is spotted, the light beam is kept in its eyes, and this appears to somewhat paralyze the victim psychologically as it does not usually flee. The gig is usually not thrown, but the fish is stabbed and pulled into the canoe. Other than the cichlids, Cuniã fishermen also take the *aruanã (Osteoglossum bicirrhosum*, Osteoglossidae) and *traíra (Hoplias malabaricus*, Erythrinidae) with gigs, among many other species that are occasionally encountered near the surface or shore at night.

Table 3.3 The common food fishes of the Rio Madeira floodplain and the months when they are most captured and the gear used to take each species.

Fish species	Months captured (J F M A M J J A S O N D)	Gear used
Cichlidae		
Tucunaré (*Cichla ocellaris*)	May–December	Gig, Bow-and-Arrow, Lure, Handline, Gillnet
Cará-Açú (*Astronotus ocellatus*)	June–December	Gig, Bow-and-Arrow, Gillnet
Characidae		
Jatuarana (*Brycon* sp.)	March–July	Gillnet, Trotline
Tambaqui (*Colossoma macropomum*)	April–August	Gillnet, Trotline
Pirapitinga (*Colossoma bidens*)	April–August	Gillnet, Trotline
Pacu Toba (*Mylossoma duriventris*)	April–June	Pole-and-Line
Pacu Branco (*Mylossoma aureus*)	April–June	Pole-and-Line
Pacu Encarnado (*Mylossoma albiscopus*)	April–June	Pole-and-Line
Piranha Caju (*Serrasalmus nattereri*)	May–October	Pole-and-Line, Gillnet
Prochilodontidae		
Curimatá (*Prochilodus nigricans*)	July	Bow-and-Arrow, Castnet, Gillnet
Jaraqui Escama Grossa (*Semaprochilodus theraponura*)	January	Gillnet, Bow-and-Arrow
Curimatidae		
Branquinha Chora (*Curimata latior*)	September–November	Castnet, Gillnet
Branquinha Cabeça Lisa (*Curimata altamazonica*)	September–November	Castnet, Gillnet
Branquinha Comum (*Curimata vittata*)	September–November	Castnet, Gillnet
Erythrinidae		
Traíra (*Hoplias malabaricus*)	May–October	Gig, Bow-and-Arrow, Gillnet
Pimelodidae		
Filhote (*Brachyplatystoma* sp.)	July–October	Gillnet
Loricariidae		
Bodó (*Plecostomus* spp.)	September–November	Castnet, Gillet
Bodó (*Pterygoplichthys* spp.)	September–November	Castnet, Gillet
Osteoglossidae		
Aruanã (*Osteoglossum bicirrhosum*)	September–November	Bow-and-Arrow, Gillnet
Pirarucu (*Arapaina gigas*)	September–November	Gillnet, Harpoon, Curumim-Line

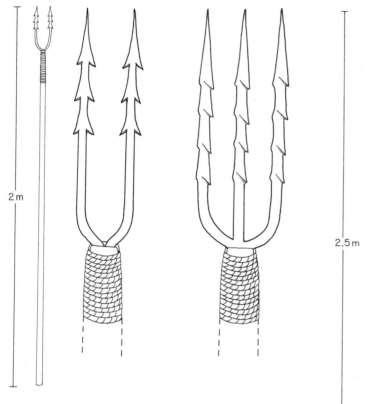

Fig. 3.15 The *zagáia*, or gig, used in floodplain fisheries.

Pindá-lure

The *pindá*, or *pindauaca*, is a type of lure attached to line-and-pole (Fig. 3.16) that is used to catch the *tucunaré* (*Cichla ocellaris*, Cichlidae). The lure is constructed from a treble hook – or three single hooks tied together – to which is tied strips of red cloth or bird's feathers. Because the *tucunaré* is only diurnally active, the *pindá* can only be used effectively during the day. Seated in the prow of his canoe, the fisherman paddles along the edge of the lake, and with one hand manipulating the pole, skids the lure back and forth across the surface of the water. Though I failed to see the resemblence, the *tucunaré* apparently thinks the red lure looks like a fish, as it violently attacks the artificial bait. Fishermen suggest, and not altogether frivolously, that the large cichlid attacks the *pindá*-lure out of spite or madness *(raiva)*. Whatever the case, it is an effective device for catching the *tucunaré* in small quantities.

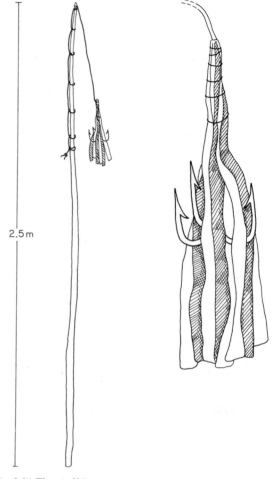

Fig. 3.16 The *pindá*-lure.

Bow-and-arrow

As water level in the Cuniã floodplain begins dropping fairly rapidly after June, the fishes in the flooded forest become more concentrated. Cuniã fishermen are some of the best piscine archers in the Rio Madeira region, and finn the bow-and-arrow (Fig. 3.17) to be useful in July and August when water transparency improves subsequent to the flood (Fig. 3.18). Although a wide variety of species is shot with the bow-and-arrow, it is the *curimatá* (*Prochilodus nigricans*, Prochilodontidae) that is most heavily attacked by the archers. This detritus feeding species is easily observed in its horizontal feeding position as it removes fine particles from submerged trunks and branches in the lakes and lagoons. Fishing arrows are constructed from the quasi-woody stems of *Gynerium*, a large grass planted or found naturally along the banks of the Rio Madeira. The arrow shaft is usually unendowed with flight feathers, and measures 1.5 to 2.0 m in length. A wooden notch is attached to the top of the arrow while a wooden, conical insertion point is attached to the other end. The one, two, or three-pronged point is made from heavy wire; the upper end of the point is fitted into a wooden head that serves as the receptacle for the arrow shaft. The arrow is made ready to fire by inserting the shaft into the point, the two parts being connected by a piece of string that, when twisted around the arrow before insertion has been made, guarantees that the fit will be tight. When the arrow is fired and sunk into a fish, the tip is usually dislodged from the shaft, but the two remain connected by the short

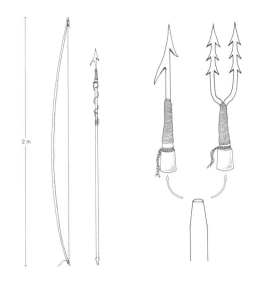

Fig. 3.17 Bow-and-arrow used by Rio Madeira fishermen.

Fig. 3.18 An archer fires an arrow at a *curimatá* (*Prochilodus nigricans*, Prochilodontidae) that is feeding on the detritus attached to the partly submerged myrtaceous shrubs along the edge of a Cuniã floodplain lake.

string. Even if the fish is not paralyzed by the arrow blast, and is able to swim away dragging the shaft behind, the fisherman easily spots his arrow because it floats.

Handline

The *currico* is a handline made from monofilament line, to the end of which is attached one hook that is baited with a piece of fish. The *currico*-handline is designed to catch predatory fishes that feed near the surface, such as the *tucunaré (Cichla ocellaris)*. To throw the *currico*, the fisherman first twirls and short piece of the line above his head or off to his side, and once the necessary momentum has been gained, the line is released to shoot to its target, which is often beneath some overhanging tree or shrub. The line is retrieved in quick, short strokes so that the bait remains on or near the surface. When the large cichlid attacks the bait, it is hooked with a quick jerk, and pulled into the canoe.

Floats

The *camurim* is a float device to which is attached a line with a baited hook (Fig. 3.19). Enough line is unwound from the float to allow the baited hook to sink to the desired depth; if catfishes are being pursued then the baited hook is allowed to sink to the bottom. The *camurim* is used mostly during the low water season when fishes are concentrated in lakes and lagoons, and a single fisherman may spread out as many as fifty of these floats. Each has but one hook and is baited with a piece of fish. When a fish is hooked and struggles, it pulls off the excess line, which is only loosely wrapped around the wooden buoy. No matter where the fish goes, the float is a telltale. At Cuniã, fishermen use the *camurim* to capture the *aruanã (Osteoglossum bicirrhosum*, Osteoglossidae). Because this fish is a surface feeder, the bated line is left dangling near the surface.

Castnet

The *tarrafa*, or castnet, is only of use during the low water period when parts of the Cuniã lakes and lagoons are less than about 2 m deep. When spread out flat on the ground, the castnet is seen to be circular; when the pulling rope, which is attached to

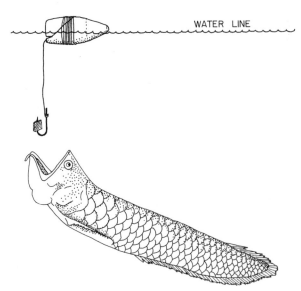

Fig. 3.19 The *camurim* float used to catch floodplain fishes.

the middle of the net, is lifted to about the shoulders of a fisherman, then the net appears conical. Fishermen usually work in pairs when castnetting, with the one in the prow of the canoe throwing the *tarrafa* and the man astern paddling. The pulling rope of the castnet is tied to one wrist, and then the upper two-thirds of the castnet is doubled or tripled and allowed to rest on the same arm to which is tied the pulling rope. A piece of the net is held off of the arm with the teeth. The free hand, palm down, is slid underneath the weighted line at the bottom of the net. The castnet is now swung backward and then thrown forward as to open in mid-air and fall on the water with its greatest circumference. The weights pull the castnet quickly to the bottom. The paddler then positions the castnetter in such a way that he is as near as possible directly above the sunken *tarrafa*. Once in this position, the pulling rope is retrieved and the weighted circumference of the opening of the net comes together on the bottom and prevents most fishes from escaping while the castnet is pulled into the canoe. Almost all commercial fishes found in the lakes can be caught with the *tarrafa*, though Cuniã fishermen seem to have best luck in capturing loricariid catfishes that live on the bottom. These medium-sized, armored siluroids often reveal their presence by air bubbles they release or stirring up hydrogen sulfide gas in the detritus.

Harpoon

The *arpão*, or harpoon, is used to kill the large *pirarucu* (*Arapaima gigas*, Osteoglossidae), an air-breathing fish that must surface every few minutes. The Amazon harpoon consists of four parts: the shaft, the steel tip, the pulling line, and the float (Fig. 3.20). The shaft ranges from about 1.5 to 2.5 m in length and is made from the corners of the large buttresses of *Mora paraensis* (Leguminosae), a heavy wood. The point of the tip is flattened but sharp, and usually has two backward projecting barbs. The top of the tip has a hollowed, conical receptacle that receives the wooden shaft. The conical part of the tip is wrapped with string in a manner that a series of loops is also formed. The pulling line is tied through these loops thus connecting it with the tip. Once the end of the shaft is inserted into the steel tip, the pulling line is stretched taut and kept in this position by a small loop that is fixed to the main line with a separate piece of string and hooks over a small nail. The pulling line is also attached to the top end of the harpoon by another piece of string, but in this case is threaded through a small loop at the top of the harpoon, thus preventing it from interfering with the throwing motion of the fisherman. Beyond the top of the harpoon there can be 20-30 m of pulling line, to the end of which is attached a large float. When a *pirarucu* is harpooned, the tip is dislodged from the shaft and the fish may pull out all of the line before the fisherman is able to manhandle it. If the line has to be released, then the large float will indicate the whereabouts of the harpooned victim.

Curumim-line

The *curumim* consists of a short line to the end of which is tied a large hook that is baited with fish (Fig. 3.21). The device is designed mostly for catching *pirarucu* (*Arapaima gigas*). The *curumim* is set in position by attaching it to the emerged top of a small tree or climbing vine in the flooded forest. The bait is kept near the surface and the excess line is placed on top of a branch or a knotch in a vine. When a *pirarucu* sucks in the bait, it is hooked, and this is often in the stomach. The small tree or vine play the fish just as would a fishing pole held by a man.

Gillnets

Gillnets have been used at Cuniã since about 1975, especially larger meshed sizes designed to catch the deep-bodied fruit eating characins, *tambaqui* (*Colossoma macropomum*) and *pirapitinga* (*Colossoma bidens*), and the *filhote* catfish (*Brachyplatystoma* sp., Pimelodidae). The abundance of the *piranha caju* (*Serrasalmus nattereri*, Characidae) discourages the use of small meshed gillnets because these fishes cut this gear into pieces when caught or when they attack entangled fishes.

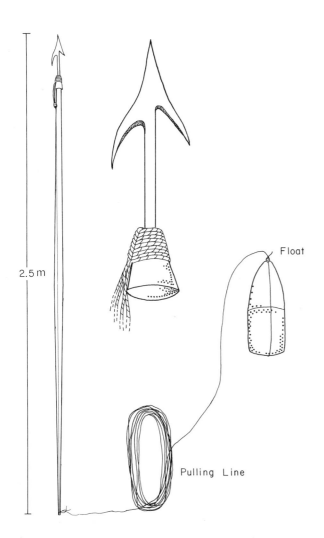

Fig. 3.20 The harpoon used to kill the *pirarucu* (*Arapaima gigas*, Osteoglossidae).

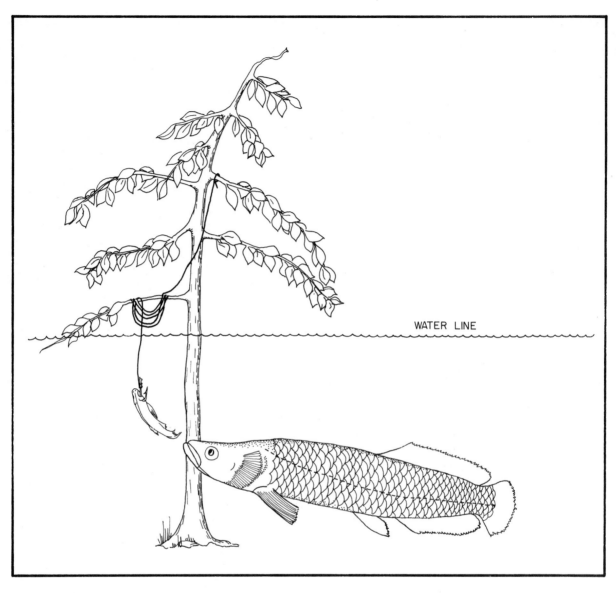

Fig. 3.21 The *curumim*-line used to catch *pirarucu* (*Arapaima gigas*, Osteoglossidae).

Lake draining

The draining of a lake is not a fishing implement per se, but it is nevertheless an effective way to make fishing easier. For the past twenty years, Cuniã fishermen have been draining a small floodplain lake during the low water season. The lake does not drain completely, but is reduced to depths of about one meter, and the fishes become much more concentrated and can be removed with castnets. The outlet to this lake is choked with aquatic herbaceous plants that serve to dam the waterbody back during the low water period. By cutting a one or two meter swath through the two kilometer long outlet (which communicates with the main lake, or Lago de Cuniã), the smaller lake can be largely drained. In 1977, I recorded three tons of fish that were removed from this small lake, *tucunaré* (*Cichla ocellaris*), and *bodó* (*Plecostomus* spp. and *Pterygoplichthys* spp., Loricariidae) being the most important.

Flooded forest fisheries

The flooded forests of the Rio Madeira valley, and indeed of all the Amazon basin, play a twofold role in relation to fisheries. Food chains based on phytoplankton and/or macrophytes are limited because of the overall nutrient poverty of the aquatic ecosystems (especially those centering on the clearwater and blackwater rivers). This places a premium on non-aquatic produced food sources that fall or wash into the waterbodies. The flooded forests are the main suppliers or this extraneous enery that fuel the food chain leading to many fish species. Amazonian fishes are adapted to eat fruits, seeds, leaves, and arthropods that fall out of the trees during the floods. Other than being an important feeding habitat for fishes, the flooded forests also offer most fish species some protection against fishermen. In the flooded forests most fish species are too widely dispersed to be caught in large numbers, and thus this habitat serves as a seasonal palisade against intensive commercial fishing.

During the floods most of the food fishes are found in the flooded forests, but the fisheries at this time of year are highly selective because only a few species can be exploited successfully on a commercial basis. The four most important genera (*Colossoma*, *Mylossoma*, *Myleus*, and *Brycon*) caught in the flooded forests of the Rio Madeira and its tributaries share an ecological common denominator, namely, they are all mainly fruit and seed eating fishes. It is this trophic preference that makes them more vulnerable than other fish taxa to commercial exploitation by man. The frugivorous fishes are adapted to go into the flooded forests and feed heavily during the few months that fruits and seeds are available to them. Their voracious appetites leave them prey to fishermen who know their feeding habits.

Pole fishing
The fishing pole used by Amazon fishermen is usually cut from the top of a small tree of the family Annonaceae. These poles are strong but flexible. Flooded forest fishermen have a very intimate knowledge of the feeding behavior of the frugivorous fishes, and know what fruits or seeds can be used as bait to catch them. Pole fishermen usually pursue the smaller species of 15-30 cm length, especially those of the genera *Mylossoma*, *Myleus*, and *Brycon*. Paddling along or through the *igapó*, the fisherman looks and listens for fishes snapping fruits off the surface of the water. The fruit eating fishes cluster beneath trees or vines where fruits and seeds are falling, and it is at these sites that the fisherman has most luck. Paddling with one hand he maneuvers his canoe into position, while with the other hand he manipulates the pole so as to whip the water with the line and fruit or seed baited hook. This action is meant to imitate fruits or seeds falling into the water, and indeed it does, as the fishes strike the bait and are then hooked and flipped into the canoe. A fisherman may catch as many as a dozen fishes from beneath one tree, before the stock is depleted or frightened away. The large fishes of the genus *Colossoma*, weighing between 5 and 20 km, can also be caught with the pole, but the method is somewhat different than described above. The tackle must be heavier and usually larger fruits and seeds are used as bait. Palm fruits (*Astrocaryum jauary*, Palmae), rubber tree seeds (*Hevea spruceana* and *Hevea brasiliensis*, Euphorbiaceae), and cucurbits (*Luffa* sp., Cucurbitaceae) are some of the favorite baits used. Once one of the large fishes is hooked, the pole is abandoned and the fisherman grabs the line and trys to pull the fish in; if the hooked fish be a large *tambaqui* (*Colossoma macropomum*, Characidae), the fisherman is often taken for a short ride through the flooded forest before the fish can be exausted and brought aboard.

The gaponga
The *gaponga* is a simple but ingenious device used to imitate the plunking of falling fruit in the water. It consists of a pole-and-line, to the end of which is tied a weight, such as a ballbearing or large screwnut. The water is gently whipped with this weight to produce the acoustic effect of a falling fruit. I have only seen it used near *jauari* trees (*Astrocaryun jauary*, Palmae); the fruits of this palm are fairly large and heavy and can be heard for some distance when they are falling into the water. As the fisher-

Fig. 3.22 The flooded forest trotline and gillnet.

man makes the plunking sound with the *gaponga*, he is ready with the harpoon in the other hand. If a large fruit eating fish surfaces to inspect the *gaponga* weight, then it may fall victim to the harpoon.

Fruit and seed baited trotline

The flooded forest trotline is constructed from a 10-50 m heavy line to which are attached hooks suspended on short lines of less than about 50 cm (Fig. 3.22). The trotline is stretched somewhat tautly between the emerged tops of inundated trees or vines. In the Rio Madeira region, the flooded forest trotlines are designed to catch mostly fishes of the genera *Colossoma* and *Brycon*. The hooks are baited with rubber tree seeds *(Hevea spruceana* and *Hevea brasiliensis)* and palm fruits *(Astrocaryum jauary)*. One of the most serious problems with trotlines is that once a fish has been hooked *piranhas* attack it. Nevertheless, the method is efficient enough to be used regularly in flooded forest fisheries.

Gillnets

The diffusion of gillnet technology has been very rapid in the last decade in the Amazon basin (Fig. 3.23). In the late 1970's, gillnets accounted for most of the catfish catch in the Rio Madeira valley; only a very small percentage of the characin and cichlid catch, however, was taken with this method. Most of the non-catfish gillnet catch in the Rio Madeira region is represented by *Colossoma macropomum* and *Colossoma bidens*, the largest characins. Petrere (1978) has also shown that most of the catch of *Colossoma macropomum* arriving in the Manaus market is largely taken with gillnets. To increase the catches of smaller species would require using smaller meshed gillnets, but these are too easily destroyed by *piranhas* and thus not very economical.

The flooded forest gillnet is usually about 10-20 m in length and 2-4 m in height; mesh sizes range from about 16-24 cm when stretched. The net is strung tautly between flexible trees or vines so that there is some give when a fish becomes entangled and fights to escape (see Fig. 3.22). The nets are usually placed near trees where fruits or seeds that the large characins feed on are falling. In the case of the large, deep-bodied characins, the gillnet has broken through the palisade of protection against large-scale fishing offered by the flooded forest to most fishes. As will be discussed later, the large, deep-bodied characins are in danger of being seriously over-exploited due both to increased gillnet and seine fishing.

Weir fisheries

The *tapagem*, or weir, used in the Rio Madeira fisheries is a net-fence or wattle placed across the debouchure of a stream or floodplain lake outlet to catch or retain fishes attempting to exit from these waterbodies (Fig. 3.24). This fishing technique was widespread among native Amerind groups of the Amazon who used plant fibers, poles, and split palm wood to construct their weirs. Cotton nets began to be used by at least the end of the nineteenth century in *caboclo* renditions of native weirs (Veríssimo 1895), but today these are replaced by nylon or monofilament gillnets or even finer meshed materials such as those used in the construction of seines.

The Rio Madeira has relatively few rainforest streams that enter directly into it, though there are numerous *furos*, or narrow channel-like escavations that cut through the high levees and connect the main river with floodplain lakes and lagoons during the floods. With rising water level, the floodplain lakes and lagoons begin to fill and eventually inundate surrounding *igapó* forest; likewise, rainforest streams become dammed back and invaded by the main river and flood laterally into the low forest. At this stage fishes begin to enter these flooded areas from the main river, and this is usually in late October and November in the upper Rio Madeira. It is still unclear to what extent the migratory characins that enter the rainforest streams and floodplain lakes invaded by Rio Madeira turbid water move out into the main river to spawn (depending on species, as mentioned earlier, sometime between late November and early February). The local residents that live along these waterbodies report that the migratory characins spawn wherever turbid water is met, and this appears to be true because schools are not observed moving through the turbid water *furos* or rainforest streams invaded by the main river. The point here is that weirs are of little use at this time of the year because there are few fish schools that transit the *furo* and rainforest stream mouths where they are placed. In the Rio Madeira region weirs are only rarely used in an attempt to catch the spawning charcins at the beginning of the floods.

When water level begins falling, the fishes are forced to migrate out of the restricted floodplain lakes and lagoons and rainforest streams where they were feeding in the previous months. Their only exit from these areas is through the narrow V-shaped cuts in the high levees through which the *furos* and and rainforest streams drain. It is at these sites that the weirs are placed. When weirs are constructed of gillnets, they block the exit of, or entangle, only those fishes comparable to or larger than the mesh size used. Most weir fishermen, however, prefer fine-meshed nets for their blockades, as these prevent the exit of all fishes. Most of the biomass of fishes moving in and out of the rainforest streams and floodplain lakes seasonally are migratory characins. By far the most important species captured is the *curimatá (Prochilodus nigricans*, Prochilodontidae). The curimatids (*Curimata* spp.) are probably as abundant, but they commanded too low of a market price in the late 1970's to interest fishermen. When schools of curimatids appear at the mouths of the *furos* and rainforest streams, the weirs are usually opened and they are allowed to flee to the main river. If water level begins to rise for a few days during the general decline after the peak of the flood, migratory characin schools often attempt to reenter the rainforest streams or floodplain *furos*, in which case the weirs are opened and they are allowed entrance, only to be caught later when they try to escape.

53

Fig. 3.23 A fisherman constructing a flooded forest gillnet.

Fig. 3.24 A fish weir of the upper Rio Madeira.

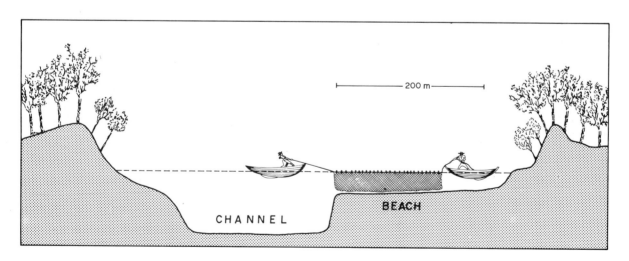

Fig. 3.25 Drifting beach-gillnet used in clearwater tributaries of Rio Madeira.

In 1977, there were about 15 weirs in the upper Rio Madeira area that were being used to supply the Porto Velho market, and also to export fish to Rio Branco, Acre. The two most productive of these yielded about ten tons of *Prochilodus nigricans* apiece. Weirs were prohibited in 1978 by Brazilian fish authorities, though they were still being used clandestinely in 1980.

Low water fishing in clearwater tributaries

Until about 1978, the large clearwater tributaries of the rightbank of the Rio Madeira where exploited commercially for fishes mostly at their mouths and to a limited extent the inundation forests were fished during the floods. As fuel prices began to spiral in the late 1970's, and catch per unit of effort began to decrease in the upper and middle Rio Madeira, Porto Velho fishermen literally invaded the clearwater tributaries, such as the Rio Jamari and Rio Machado, during the low water period. Prior to 1977, Rio Madeira fishermen did not have an efficient method for catching fishes on a large scale in the clearwater rivers. During the day, scincs are ineffective in the clearwaters because the fishes can see them, and even at night when they might be more useful, the large black *piranha (Serrasalmus rhombeus*, Characidae) is such a menace that fishermen are reluctant to employ them lest they be lacerated by these sharp-toothed predators.

In 1977, commercial fishermen began experimenting with 100-300 m long monofilament gillnets that could be drifted downstream with the current and across shallow beach areas at night (Fig. 3.25), when many fish taxa, such as *Mylossoma*, *Myleus*, *Semaprochilodus*, and *Brycon*, become concentrated in these habitats. These fishes, which are among the most abundant species in the clearwater rivers, appear to seek the beaches at night to avoid heavy predation by the large catfishes, a few of which are usually caught along with the characins with the drifting beach-gillnets. Even when the characins are frightened at night, they are reluctant to leave the beaches and thus fishermen can often exploit the same school two or three times.

The new frontier:
The Rio Mamoré and Rio Guaporé

The fisheries frontier of the Rio Madeira valley began to shift south, above the rapids and into the Rio Mamoré and Rio Guaporé, after about 1977 when it became obvious to some commercial fishermen that the Rio Madeira itself had already been exploited to its maximum limit and that catch per unit of effort would decline henceforth. Refrigeration companies from Porto Velho paid professional fishermen to reconnaissance the Rio Mamoré and Rio Guaporé basins, and these initial surveys suggested – and correctly – that there were sufficient stocks of fishes to be exploited commercially on a large scale, and furthermore, that the local fisheries of the region were little developed because of the availability of beef from Bolivia. By 1978, boats with freezing units were plying up the shallow Rio Mamoré and Rio Guaporé and bringing back capacity catches. Bolivian fishermen were also engaged by Brazilian companies to exploit the Rio Mamoré, and then export their catches across the border to Guajará-Mirim, Brazil.

In the late 1970's, the commercial fisheries of the Rio Mamoré and Rio Guaporé were mostly directed toward a few species that could be exported to Porto Velho or Southern Brazil. Catfishes were being caught mostly with drifting deepwater-gillnets. The most important species captured in 1978 and 1979 were *Pseudoplatystoma tigrinum*, *Pseudoplatystoma fasciatum*, *Brachyplatystoma flavicans*, and *Brachyplatystoma filamentosum* (Pimelodidae). Scaled fishes exported from the Rio Mamoré and Rio Guaporé consisted almost entirely of *tucunaré (Cichla ocellaris*, Cichlidae), *tambaqui (Colossoma macropomum*, Characidae), *curimatá (Prochilodus nigricans*, Prochilodontidae), and *jatuarana (Brycon* sp., Characidae). In the late 1970's, *Cichla ocellaris* was taken mostly with gig at night in the swampy areas along the Rio Guaporé. Most of the *Colossoma macropomum* catch came from the Rio Mamoré, where it was captured with gillnets by Bolivian fishermen and then transported downstream in crude boats equipped with ice boxes and then sold to Brazilian refrigeration companies (Fig. 3.26). A large *Brycon* (I measured market specimens

Fig. 3.26 Crude fishing boat used by Bolivian fishermen exploiting the Rio Mamoré (in Bolivia) and selling catches to Brazilian refrigeration companies in Guajará-Mirim.

Fig. 3.27 The *calhapo*, or underwater pen, used to transport live fishes and turtles in the Rio Mamoré and Rio Guaporé region.

of at least 80 cm standard length) was being catured at the Cachoeira da Esperanza, the cataract about 20 km up the Rio Beni. This beautiful fish was exploited mostly during the low water season when it was migrating upstream.

Bolivian and Brazilian fishermen report that the *curimatá (Prochilodus nigricans*, Prochilodontidae) is the most abundant food fish in the Rio Guaporé and Rio Mamoré. A quite similar species, *Prochilodus platensis*, has also been reported to be the most abundant fish in the Rio Pilcomayo of Eastern Bolivia (Bayley 1973). Unlike the Rio Pilcomayo area, however, fishes of the genus *Prochilodus* in the Rio Guaporé and Rio Mamoré region are considered fit only for consumption if nothing else can be found. The pejorative name of the *curimatá (Prochilodus nigricans)*, in the Rio Mamoré area, is '*quebra galho*', which may be translated loosely as 'it will do in a pinch'. For at least two decades, however, Guajará-Mirim fishermen and fishmongers have been exporting *Prochilodus nigricans* to Rio Branco, Acre; before the road connection was made in the late 1960's, the fish were transported in DC-3 aircraft, whereas recently they are trucked in a two day journcy. Although Rio Mamoré and Rio Guaporé fishermen did not have ice until 1978, they were still able to catch and preserve large quantities of fish. This was done in a device called the *calhapo*, which is an underwater holding pen for fishes and turtles (Fig. 3.27). The cages that I measured in Guajará-Mirim were about 15 m long, 4 m wide, and 2.5 m deep. The *calhapo* is towed along side or behind a boat, and when a catch has been made the fishes are dumped alive into the underwater cage. Fishermen told me that they have left fishes in *calhapos* for as long as two weeks, but less than a week is best because mortality is always high, and fifty percent losses are accepted as normal. Although the *calhapos* were being replaced by ice boats in 1978, they were still commonly seen along the waterfronts of Guajará-Mirim (Brazil) and Guayará-mirin (Bolivia). They were being used both for transporting fishes and turtles. The Brazilian government has prohibited the capture of turtles or their importation from Bolivia, but chelonian prices and demands are astronomical, and thus a brisk trade goes on across the border, especially at night when the reptiles are ferried across the Rio Mamoré in small boats. In Porto Velho a large river turtle *(Prodocnemis expansa*, Pelomedusidae), depending on its size, was fetching the equivalent of $50-100 in 1978.

CHAPTER 4

Fishing area, effort, and yield

Fisheries regions of the Rio Madeira drainage system

The drainage system centering on the Rio Madeira can be divided into three main commercial fishing regions:

1) Beginning upriver, the first is the large area embracing the Rio Beni, Rio Mamoré, and Rio Guaporé. The Rio Beni lies entirely within Bolivia, while the Rio Mamoré and Rio Guaporé are shared with Brazil. Large-scale commercial fishing began in this region only in the late 1970's when Brazilian refrigeration companies began buying fishes and exporting them to Southern and Southeastern Brazil, Porto Velho (Rondônia), and Rio Branco (Acre). The Brazilian city of Guajará-Mirim on the right bank of the Rio Mamoré serves as the main fisheries port of the region, including nearby Eastern Bolivia. In 1979, at least 700 tons of fish were exported through Guajará-Mirim.

2) In total area the largest fisheries region of the Rio Madeira is that dominated by the Porto Velho fishing fleet, and this embraces the area of the rapids in the upper course of the river to just below the Rio Aripuanã. The largest annual catch from this region was about 1,800 tons in 1974, but yields have declined considerably in recent years (see below). This is the region that will be discussed in detail below.

3) The lower course of the Rio Madeira is, for all practical purposes, part of the Central Amazon fisheries region, serving mostly Manaus, but also smaller cities such as Itacoatiara. The lower Rio Madeira lies closer to Manaus (on the Rio Negro) and Itacoatiara (on the middle Rio Amazonas) than it does to Porto Velho. In 1977, at least 2,400 tons of fish captured in the lower Rio Madeira entered the Manaus market (Bayley 1978); the Itacoatiara fleet also exploited the lower Rio Madeira but its catches were not recorded.

Yields by species

Although the Amazon basin has the world's most diverse freshwater fish fauna, there are relatively few species that can be exploited for food on a large scale. In the case of the Rio Madeira, I believe that all of the potentially important species are already being exploited, and that no new food fish discoveries will be made. This observation is based both on data from the commercial catches and from intensive and extensive experimental fishing in the system with gillnets and seines. Any significant increases in yields will be based on the species that are at present being exploited.

To give a better idea of the most important food fish species of the Rio Madeira region, I have combined the annual catches of 1977, 1978, and 1979, and then calculated the percentage of the total catch that each species, or group of closely related species, represents (Table 4.1). Nine genera, including about 18 species, accounted for 87 percent of the total three year catch. About five species, however, represented 70 percent of the total.

Table 4.1 Total 1977-1979 catch of Rio Madeira food fishes. For comparison, the relative importance of each species or group of closely related species is given for the Manaus market for 1976. Manaus data from Petrere (1978b).

Food fish species	Porto Velho, 1977-1979 total catch (in tons)	%	Relative importance in Manaus catch, 1976
1. **Dourada** *Brachyplatystoma flavicans*, Pimelodidae	571	21	?
2. **Jatuarana** *Brycon* sp., Characidae	427	16	Seventh
3. **Curimatá** *Prochilodus nigricans*, Prochilodontidae	352	13	Third
4. **Pacu** *Mylossoma duriventris* *Mylossoma aureus* *Mylossoma albiscopus*, Characidae	282	10	Fourth
5. **Jaraqui** *Semaprochilodus theraponura* *Semaprochilodus taeniurus*, Prochilodontidae	274	10	Second
6. **Tambaqui** *Colossoma macropomum*, Characidae	117	04	First
7. **Sardinha** *Triportheus angulatus* *Triportheus elongatus* *Triportheus* sp., Characidae	98	04	Ninth
8. **Branquinha and Cascudinha** *Curimata latior* *Curimata altamazonica* *Curimata amazonica* *Curimata vittata*, Curimatidae	72	03	Tenth
9. **Tucunaré** *Cichla ocellaris*, Cichlidae	67	03	Eighth
10. **Pirapitinga** *Colossoma bidens*, Characidae	59	02	Fifth
Sub-total (1-10)	2318	87	
All other species	356	13	
Total	2675	100	

Though the order of the species may change in coming years, it is highly unlikely that any species not listed will move into the top five. It is probable that detritus feeding fishes of the genera *Gurimata* and *Prochilodus* will move up in importance as the more favored species become scarcer.

A comparison of the important food fishes of the Porto Velho and Manaus markets shows that the same species, or closely related species, are of importance in the respective cities. The only exception is that in the Manaus market the anostomids of the genera *Schizodon* and *Rhytiodus* are in the top ten taxa, but were of only minimal importance in the Porto Velho catch in the late 1970's. Almost no data on catfish catches was collected for most of the state of Amazonas in the 1970's; most of the siluroid catch from this region was exported to Bogotá, Colombia and Southern and Southeastern Brazil. Bayley (1978) estimated that at least 11,500 tons of large catfish were caught in the state of Amazonas in 1977, and it is known that the *dourada* (*Brachyplatystoma flavicans*, Pimelodidae), the most important food fish species of the middle and upper Rio Madeira, was also the most important siluroid species being exploited in the state of Amazonas. There is almost no difference, then, in the species of fishes that are exploited in the Rio Madeira and those arriving in the Manaus market from a large area in the state of Amazonas.

History of annual catches

The Rio Madeira fisheries, based on the Porto Velho fleet, peaked in total annual yield in 1974 when there was a steep increase in fishing effort and extremely low water levels (see below). In 1974, the Porto Velho fleet – including refrigeration companies that were exploiting large catfishes – brought to market an estimated 1,800 tons of fish, but since that peak-year yields have declined by about one-half (Figs. 4.1 and 4.2). Rio Madeira fishermen, many of whom are excellent observers and have worked in the fisheries for twenty years or more, believe that the principal *environmental* factor affecting total annual catch is water level. Their belief is worth examining in some detail in light of the evidence that is available.

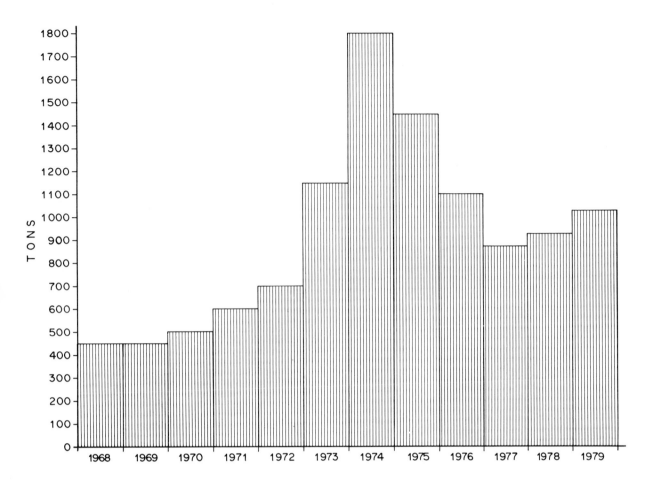

Fig. 4.1 Total annual catches of the Porto Velho fishing fleet of the Rio Madeira.

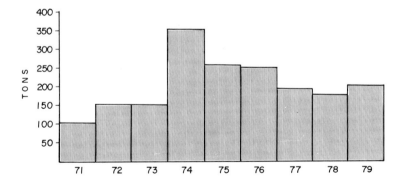

Fig. 4.2 Total annual catches of the Teotônio cataract fisheries. These include almost entirely catfishes.

Mimimum low water levels

The years for which Rio Madeira hydrological data are available (1974-1979) show that the minimum low water level recorded was in 1974, and this was about four meters below that of 1977, which was the highest recorded for the period (see Fig. 1.8). Although Porto Velho fishermen and refrigeration companies did not anticipate the extremely low water levels in 1974, they almost coincidentally began to increase their fishing effort in 1973 and were handsomely rewarded for this in the following year when there were unusually large upstream migrations of fishes (riparian folk, fishermen, and catch data verify this). In 1977, there was a large flood and the subsequent low water levels remained relatively high, and the upstream fish migrations were smaller than in the previous three years. The 1977 catch was the lowest of the six years of the period 1974-1979. In 1978 and 1979, the Rio Madeira catch data indicate that there were larger upstream migrations during the low water period than in 1977, but not on the scale of those of 1974. Only another extremely low water year such as 1974 would make it possible to know if fishermen could catch a comparable amount of fish, or if, as I believe, the commercial fish biomass has already been substantially reduced in the Rio Madeira region and that another annual yield by the Porto Velho fleet on the order of 1,800 tons is highly unlikely.

The largest part of the fish biomass involved in the upstream, or *piracema*, migrations is made up of non-predatory characins (see Chapters 3 and 5). It is not known for certain why the intensity of their upstream migrations is positively correlated with minimum low water levels, but one factor suggested to me is predation. As mentioned earlier, much if not most of the characin food fish biomass of the Rio Madeira basin is thought to reside during most of the year in the large clearwater tributaries of the rightbank. It is known that the migratory characins migrate from one tributary to another farther upstream, and most species do this during the low water period. These movements are thought to be dispersal migrations in which fishes are recruited from the lower reaches of the system where most of the nursery habitats are found (Goulding 1980; and see Chapter 5 in present work). I believe that the preference of the migratory characins for the clearwater tributaries during the low water period has evolved as an adaptation – through natural selection – to decrease predation, especially by the large catfishes.* The migratory characins have relatively large eyes, and their visual sense undoubtedly helps them escape predators better in clear than in turbid water in most years. I say most, because the clearwater tributaries of the Rio Madeira, in extremely low water years, are so reduced in volume and area that it would appear that the prey fishes (of which the characins undoubtedly represent a large part of the total biomass) would have an extremely hard time escaping predators, in spite of the relatively good transparency. When water level gets too low, the migratory characins move downstream and into the Rio Madeira, and these movements are probably so similar to dispersal migrations that they cannot be distinguished from them. In a year such as 1974, then, the large catfishes literally chase the prey fishes – through predation – out of the clearwater tributaries and into the Rio Madeira, which is deeper and larger than its affluents. Commercial fishing, however, presents a new form of predation, evolutionarily speaking, and the migratory characins fleeing to the Rio Madeira from its clearwater tributaries in extremely low water years, may now suffer greater losses from the seine than from the siluroid.

Maximum high water levels

Rio Madeira fishermen believe that large floods have two principal effects on the fisheries, which are not necessarily evidenced in the catches of the years with larger than normal inundations. The first is referred to as a 'releasing' (*o peixe é soltado*) of fishes from lakes and lagoons on the high part of the floodplain, especially in the lower course of the river. Many floodplain lakes and lagoons only communicate with the main river in years when there are large floods, and in the intervening years, the fishes remain imprisoned in these waterbodies. With the large floods, the migratory fishes are released and, depending on the subsequent low water period, they greatly add to the total catch. I believe that the quantity of imprisoned adult fish

*As mentioned earlier, adult migratory characins in the Rio Madeira basin are highly dependent on flooded forests for their food, and this fact explains their presence in the tributaries during the floods. During low water, however, food is scarce for non-predatory fishes, and thus the reasons why the migratory characins prefer the clearwater tributaries over the Rio Madeira must be explained by other factors. Predation is hypothesized to be the main factor.

suggested by fishermen is greatly ee;gerated, but that there may be a significant number of young fish that become trapped in these waterbodies, and that, when released with a flood comparable to one that allowed them to get to these areas in the first place, they may add significantly to recruitment into the fisheries. If the young fish are trapped too long in these waterbodies, a combination of predation and limited food supply probably reduces their numbers to insignificance in terms of potential recruitment into the fisheries. Since we do not know how extensive these landlocked waterbodies are, and the nature of the fish faunas (especially alevins) in them, we cannot yet judge their potential importance to the fisheries of the Rio Madeira region.

Fishermen also believe that large floods produce (*criar*) more fish, especially in the lower reaches of the system where floodplains are largest and where most nursery habitats are likely to be. They suggest that two or three years after a large flood catches will be better than normal as the successful year-class is recruited into the fisheries. A large flood might lead to an especially successful year-class because young fish mortality would be reduced as a consequence of greater food availability (probably zooplankton) due to greater area flooded and longer flooding times, and perhaps predation pressure would even be decreased on young fish that are able to spread out over a wider area during the first few months of life (the time when they might be most vulnerable to predation). This is a reasonable hypothesis, but biomass and mortality studies embracing years which included large and small floods and minimum and maximum low water periods, would be needed to test it statistically. In closing, it should be noted that positive correlations have been made between flood height and total catch in African river fisheries and elsewhere (Welcomme 1979).

Seasonality in catches

The putative stereotype of the non-seasonal tropics would find little support among Rio Madeira fishermen, whose economical livelihood is severely cut back during the annual 'winter' of flooding. The 10-13.5 m annual fluctuation in water level of the Rio Madeira is the most important factor affecting the temporal distribution of catches. The fisheries data between 1977 and 1979 clearly show that food fish behavior is predictable, and consequently, the montly distribution of catches follow about the same pattern from year to year (Fig. 4.3).

The low water period (from about July through November) accounts for the largest part of the annual catch. This is when migratory characins are captured at the mouths of the tributaries or moving upstream (*piracema* migrations) in the Rio Madeira. Nearly all of these fishes are captured with seines. The second most productive period of fishing is between about mid-November and the end of December when water level begins to rise rapidly and the migratory characins descend the rightbank tributaries to spawn in the turbid Rio Madeira. These schools are captured in the rivermouths of the affluents before they have a chance to enter the Rio Madeira. During the floods fishing is very difficult because the fishes are widely dispersed in the flooded forests, and no more than about five percent of the total annual catch is taken in this habitat. During the high water period in all three years for which monthly data were recorded, relatively large catches of fishes of the genus *Semaprochilodus* were made in late March or April when these detritivores were caught migrating from one rightbank tributary to another farther upstream. This is the only significant fishery of the high water period.

Catch per unit of effort

In the evaluation of a fisheries, total catch alone means very little in terms of gaining insights into sustained yields, or the quantity of marketable fishes that can be harvested, on the average, from year to year over a long period of time without destroying the stocks through over-exploitation. Total annual catches only become meaningful, from an ecological point of view, when they can be correlated with fishing effort, and ideally, with environmental conditions over a given period of time. Fishing effort can be measured in a variety of

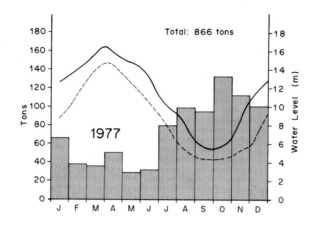
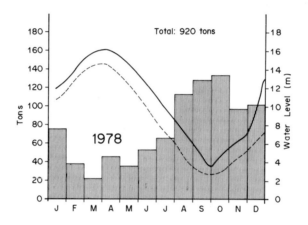
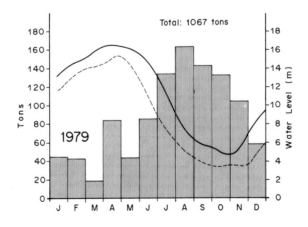

Fig. 4.3 Monthly catches of the Porto Velho fishing fleet in 1977, 1978, and 1979 in relation to water level. Monthly high and low water levels are represented respectively by the solid and broken lines. Note that the monthly distribution of catches is about the same in all three years.

ways, but almost always includes some measurement of the amount of time fished, based on men or gear or both. In this study a man-day is defined as whatever period, however long or short in the course of a day, that an individual fisherman spent commercially exploiting fish. This includes every day between the time that he left port and his return with a catch. Most fishermen work in groups on boats, and hence total man-days were calculated collectively for each fishing trip. Cooks were also included in this calculation as they often participate in fishing operations as well. Thus, for example, if a boat with ten men (including cook) was out for ten days, then a total of 100 man-days of fishing was registered. The number of times that fishing gear is used (e.g., the number of times that the seines are payed) is an alternative to using man-days as it might ideally show more accurately how much time was actually spent fishing. Petrere (1978a), however, has convincingly shown for the fisheries of the Manaus market that man-days (though not necessarily in the same way that I measured them) give a more accurate estimate of fishing effort than do estimates of time that gear was used (or reported used). This is apparently due to the inability of fishermen to accurately report the number of times that they used each type of gear and to the capricious manner in which they often fish. Other than man-days, we also registered the amount of diesel oil, gasoline, and lubricant – the latter two converted into liters of diesel oil based on price – that were spent in fishing operations in 1977; from this we were able to calculate the fish kilogram yield per liter of fuel used.

Catch per unit of effort in terms of man-days

The year 1977 was used as a base-line to determine the total fishing effort of the Rio Madeira region above the Rio Aripuanã. Information supplied by the Porto Velho fishermens' cooperative, refrigeration companies, and our own data collectors since 1977, indicate that total fishing effort of the Rio Madeira region remained about the same between 1974 and 1979. Thus the data collected on total man-days fished in 1977 can be used with some confidence to extrapolate catch per unit of effort

between 1974 and 1979, the period for which total annual catch data are available. In 1977, we registered about 33,656 man-days of fishing for the commercial fisheries of the Rio Madeira above the Rio Ariquanã, with at least 250 fishermen involved in this total effort. With the 1977 man-day total used as a base, then mean yields ranged from a high in 1974 of 53.9 kg/man-day to a low in 1977 of 25.7 kg/man-day fished (Fig. 4.4). Man-day yields improved slightly in 1978 and 1979 over the 1977 low, but they were still at least 20 kg below the 1974 high.

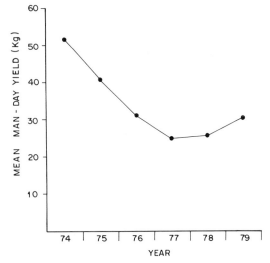

Fig. 4.4 Catch per unit of effort of the Rio Madeira fisheries expressed in mean man-day yield (kg) for the period 1974–1979.

Man-day yields are not yet available for the Manaus market, the largest in the western Amazon, but Smith (1979) estimated hourly yields of fishermen in the Itacoatiara area (near the mouth of the Rio Madeira) of the Rio Amazonas in 1978. For comparative purposes, I have converted his hourly yields into mean man-day yields, based on his statement that Itacoatiara fishermen work an average of about six hours per day. This gives 22.8 kg/man-day for the Rio Amazonas fisheries near Itacoatiara. This is surprisingly close to the 27.3 kg/man-day yield for the Rio Madeira fisheries in 1978. The nearly comparable man-day yields do not indicate that the two areas, the Rio Madeira and Rio Amazonas, are comparably productive in fishes. The Rio Amazonas, with its much larger floodplain, is certainly much more productive, in

65

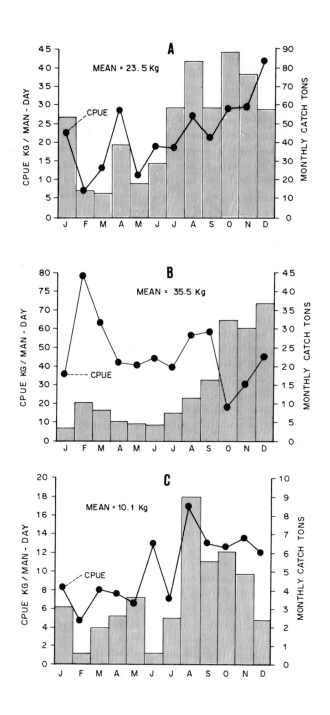

Fig. 4.5 Catch per unit of effort (CPUE) in terms of mean monthly kilogram yield per man-day fished for different types of fisheries of the Porto Velho fleet of the Rio Madeira in 1977. A) Seine fisheries. Note that there is a general correlation between total monthly catch and catch per unit of effort. B) Teotônio cataract fisheries. Note that there is no general correlation between monthly catches and catch per unit of effort. This is due to the fact that Teotônio is a geographically fixed fishery, and when fish migrations begin large numbers of fishermen are attracted to the rapids, and greatly add to the total effort but not proportionately to the total catch. C) The Cuniã floodplain fishery of the Rio Madeira. Note that there is a general correlation between monthly catches and catch per unit of effort, with the low water season being the most productive period.

terms of total catch per given length of river, than is the Rio Madeira (Bayley 1978). The town of Itacoatiara had about 30,000 inhabitants in 1978, and was largely dependent on fish for its animal protein supply, whereas Porto Velho had a population of a least 110,000 in the same year and was largely dependent on beef and not the local fisheries. The large floodplain area of the middle Rio Amazonas, and the local demand for fish, allowed Itacoatiara to support at least eight times, relative to population, the number of fishermen as Porto Velho. The fact that the Rio Madeira and middle Rio Amazonas man-day yields are so close leads me to believe that, given the present economy of Amazon fisheries, that somewhere between about 20 and 30 kg per man-day fished is the minimum level at which the commercial fisheries can function economically; below that level I would expect the total number of fishermen to decrease, or the price of fish to increase. It will be important to see if the large Manaus market turns in similar results.

Above were compared the annual mean man-day yields, and we may now turn to the per diem catch per unit of effort for the different types of fisheries of the Rio Madeira (based on 1977 data). Four types of fisheries will be considered here: cataract fisheries mostly for large catfishes; seine fisheries mostly for migratory characins; Rio Madeira floodplain fisheries based mostly on tranditional gear, and; flooded forest fisheries. The highest man-day yields of the Rio Madeira fisheries in 1977 were turned in at the Teotônio cataract (Fig. 4.5); the mean was 35.5 kg/man-day fished, representing about 21.5 percent of the total Rio Madeira catch studied. The relatively high man-day yields at the rapids were due to the concentration of upstream migrating catfishes near the cataracts and to the fact that fishermen exploiting this area spend almost no time traveling. In terms of total catch, the seine is the most important gear of the Rio Madeira fisheries, representing about 70 percent of the total catch; the mean yield based mostly on the seine was about 23.5 kg/man-day. The Rio Madeira floodplain fisheries represented only about five percent of the total catch, but in terms of mean man-day yields are instructive because they were based mostly on traditional gear such as gigs and bows-and-arrows. The mean yield of these traditional floodplain fisheries was only 10.1 kg/man-day fished; the fishermen involved in the Cuniã floodplain fisheries, however, are also subsistance farmers and thus able to sustain themselves on man-day yields well below those of the other commercial fisheries of the region. The flooded forest fisheries were the least productive, both in terms of total catch and man-day yields. Representing only two percent of the total catch, the mean yield was likewise minimal at 7.5 kg/man-day fished.

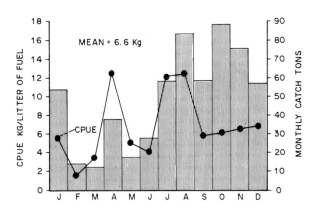

Fig. 4.6 Catch per unit of effort (CPUE) in terms of kilogram yield per liter of fuel spent in 1977. The highest yields per liter of fuel spent are in April, July, and August when fishermen wait in the tributary mouths for the migratory characins to descend the affluents. Later on in the low water period, fishermen do more traveling up-and-down the Rio Madeira in search of *piracema* schools (see Chapter 3), and thus their yields per liter of fuel spent are lower.

Catch per unit of effort in terms of fuel

There are no comparative data available with which the Rio Madeira fisheries might be compared in terms of yields per liter of fuel spent. The 1977 data, then, are offered here mostly to give an idea of the energy costs of the Rio Madeira fisheries, and later they can be compared with other areas as data become available. The mean yield in 1977 was 6.6 kg/liter of fuel spent (Fig. 4.6). If this is converted into relative prices at the time, then the ratio is about 7.1.

Catch per unit of effort in terms of ice

It takes energy to make ice, and in the case of the Rio Madeira region this energy is diesel oil that fuels the Porto Velho thermoelectric power plant that supplies the ice companies. I was unable to calculate the energy that goes into making ice in the Porto Velho area, but in 1977 we registered the amount of ice that was used in the commercial fisheries. About 2.1 kilos of block ice were used for each kilogram of fish delivered to market (calculated only from those fishing operations using ice).

CHAPTER 5

Natural history of the flood fishes

Ideally the fish fauna of a river system should be investigated from the three principal levels of biological organization – the species, the community, and the ecosystem – if its natural history is to be understood in the broadest terms. The high diversity of the Amazon fish fauna, and lack of sound taxonomical studies for most groups, make the study of communities very difficult in the present state of knowledge. My own approach has been to take a binocular perspective by concentrating on the specific and ecosystem levels, and extrapolating community structure, in a cautious manner, from these polar viewpoints. As for species, the food fishes are an excellent point of departure because of their abundance – hence importance in the ecosystem – and a constant and intensive fishing effort directed towards them helps reveal many ecological patterns, and especially migratory behavior. Before discussing each of the food fish species in detail, I will first present an overview of what is known about their migratory and feeding behavior in relation to the Rio Madeira as a system.

Fish migrations in the Rio Madeira basin

Most of the larger characin taxa (*Colossoma, Brycon, Mylossoma, Triportheus, Leporinus, Schizodon, Rhytiodus, Prochilodus, Semaprochilodus, Curimata, Hemiodus,* and *Anodus*) of commercial importance are represented by species that are migratory fishes in the sense of forming large schools and migrating in the rivers at some time of the year. The fisheries for these migratory fishes were discussed in detail in Chapter 3, and it was shown in Chapter 4 that they account for a major part of the total catch of the Rio Madeira.

The main environmental factor influencing characin migrations in the Rio Madeira basin is river level. As already discussed, two distinct migrations are detected annually, namely, one that lasts from about the beginning to the middle of the floods when the migratory characins descend the affluents to spawn in the Rio Madeira, and the second, lasting from about the height of the flood throughout the low water period, when the same species again move down the affluents (which they re-entered subsequent to spawning) and then upstream in the Rio Madeira until choosing another tributary.

The first fishes to descend the tributaries to spawn in the Rio Madeira are *Semaprochilodus theraponura* and *S. taeniurus* (Prochilodontidae), and this is usually in late November and early December. By mid-December many other species are descending the affluents and, depending on the species, these migrations last until at least mid-February, though for most species the spawning period appears to end before about mid-January. It is not known how far, upstream or downstream, the schools of migrating characins go to breed after entering the Rio Madeira, but fishermen suggest that it is close to the confluence of the tributary and the principal river (say within several kilometers). Once entering the Rio Madeira, the characin schools either dive into deeper water or disperse because they are no longer seen by fishermen. Fishes of the genera *Semaprochilodus* and *Prochilodus*, however, reveal their presence along the confluence area by the sounds that they make, and thus fishermen are led to believe that other taxa are also spawning near or just a few kilometers downstream of the affluent mouth, that is, in the area where the two water types are mixing. Fishermen also observe and catch small groups of spawned characins moving into and upstream in the tributaries, and thus they believe that the fishes, subsequent to breeding, return to the same affluent they migrated out of.

By the end of January, most of the spawned characins are in the flooded forests of the tributaries. Frugivorous taxa such as *Mylossoma*, *Triportheus*, and *Colossoma* are easily observed in these flooded forests feeding on fruits and seeds. Fisheries data, our own gillnet surveys, and reports of local residents, strongly suggest that most of the fish biomass of the middle and upper Rio Madeira region is found in the large rightbank clearwater tributaries (e.g. Rio Aripuanã, Rio Machado, and Rio Jamari) during the floods. The migratory characins are dependent on the flooded forests for most of their food, and since the tributary floodplains are inundated sooner than the floodplain of the Rio Madeira itself, and also for a longer period of time, it is to the affluents that the fishes go to find their precious sustenance.

Most of the large rightbank tributaries of the Rio Madeira have flooded forest for about five to six months per year, or mid-December to mid-June, depending on river levels in the particular year. The first migratory characins observed to leave the flooded forests and form schools and move down the tributary to the principal river are detritivores of the genus *Semaprochilodus*, and this is usually near the height of the flood; the exploitation of these fishes was discussed in Chapter 3. As river level continues to fall, and the flooded forests are drained, other taxa begin descending the affluents, and these migrations continue, with various species involved at different times, throughout the low water period. The fishes that are captured soon after leaving the flooded forests (depending on the species, from about mid-March to mid-July) are referred to by fishermen as *peixe-gordo*, or fat fish, an allusion to the large fat stores that the characins have built up while feeding in the flooded forests. By August and September, when many fish schools are encountered moving upstream in the Rio Madeira, the migrations in general are called the *piracema* (see Chapter 3).

Once characin schools enter the Rio Madeira from its tributaries, they are only observed to migrate upstream. Fishermen often accompany these schools for several days in an attempt to catch them. When pursued, the fishes are said to move more at night and thus are more difficult to catch. One wonders, in fact, to what extent commercial fishing affects the migratory behavior of these characins. Eventually the schools enter a tributary, or less so, an adjacent floodplain area of the Rio Madeira, farther upstream. The present data, however, do not suggest what factors might determine what affluent is chosen by the migrating fishes, or how far up the Rio Madeira itself that they migrate in any particular year.

Rio Madeira fishermen believe that the main nursery area for the fishes that they exploit is in the lower reaches of the river near the confluence with the Rio Amazonas. The floodplain is much larger there and flooding times are longer because the Rio Amazonas dams back the Rio Madeira. The lower Rio Madeira floodplain region is also studded with numerous lakes and lagoons and the alevins of the migratory characins are known to be abundant in these waterbodies. The alevins are also found in the floodplain lakes and lagoons of the upper and middle Rio Madeira, but these waterbodies are much more restricted in size than those of the lower reaches of the system. Most recruitment into the fisheries, then, probably originates in the lower reaches of the system. The migratory fishes are thought to take two or three years to migrate from the lower part of the system to the upper Rio Madeira, or as far as the Rio Machado. The *piracema* migration, then, is not a singular event, completed in one year, but a movement from one tributary to another farther upstream in the course of several years. To reiterate, the fishes are not thought to return back downstream, and this demands some further explanation if a reasonable hypothesis is to be presented of the life cycle of the migratory characins in the Rio Madeira basin.

Since 1977 we have made extensive fish collections in the middle and upper Rio Madeira basin, and these have clearly shown that the alevins of the migratory characins are very rare or entirely absent in the tributaries of the principal river. The migratory characins in the triburaties are mostly adult fishes, though a few species may be represented by sub-adults as well. The alevins of these fishes are found in the floodplain lakes and lagoons of the Rio Madeira. The limnology of Rio Madeira floodplain waterbodies has not been studied in any detail, but it appears to be very similar to that of the Rio Solimões which has been investigated more thoroughly. Like the waterbodies of the Rio Solimões, those of the Rio Madeira are characterized by extensive macrophyte development (the floating meadows) and seasonal plankton blooms. The floating meadow biotope has been shown by Junk (1973) to be extremely rich in invertebrate animals in the root zone, and Bayley (1979), who has been studying young fish in Rio Solimões waterbodies, has emphasized the potential importance of these habitats in transferring energy to alevins. Schmidt (1971a, 1973b) has also shown that the best plankton production in Amazon waterbodies is on the floodplains of the turbid water rivers where there is an annual injection of nutrients. The turbid water, then, supplies the necessary nutrients to build up a food chain for the alevins of the fishes with the largest biomasses. Nutrient poor waters, such as those of the clearwater tributaries of the Rio Madeira, are unable to support extensive macrophyte communities and plankton production also appears to be very limited. Though older fishes can feed in the nutrient poor tributaries, because of the extensive flooded forests where they find fruits, seeds, arthropods, and detritus, the young fishes must go to or be placed by spawners in, habitats where in situ production (i.e. plankton and macrophytes) is much higher. I believe that this is the reason that the migratory characins descend the tributaries to spawn in the Rio Madeira.

What is least clear, however, is the destiny of fry soon after birth. Two possibilities, however, suggest themselves. First, that there is a general downstream displacement of young fish to the lower course of the river soon after birth. This would place them where nursery habitats are most extensive. Or second, there is no significant downstream displacement, but that young fry try to enter the floodplain areas that are nearby the place of their birth. Since these are limited in the middle and upper Rio Madeira region, one would then expect extremely high mortality in young fish populations looking for - but largely not finding - appropriate nursery habitats on the limited floodplain. If this is the case, then most of the migratory characins that move up the Rio Madeira are born in the lower course of the system, or perhaps even in the Rio Amazonas. With a large floodplain area nearby, the fishes born in this area have a much higher survival rate, and thus can be recruited into the middle and upper Rio Madeira where mortality of young fish is too high to meet the potential for recruitment. The upstream migrating characins do not fill up, or over-populate, the upper Rio Madeira system, including the tributaries into which they move, because they are only replacing natural mortality in a region that is unable itself to supply the necessary recruitees.

Catfish migrations

In contrast to migratory characins, most catfishes travel near the bottom and thus their seasonal movements are much more difficult to detect. For studying siluroids, however, the Rio Madeira has one great advantage over the Rio Solimões-Amazonas where catfishes are undoubtedly more abundant because of the higher productivity of that system. This advantage is the stretch of cataracts - and especially the Cachoeira do Teotônio - in the upper course of the Rio Madeira. The catfishes must fight their way through the turbulent rapids, and, in so doing, are easily detected. Intensive commercial fishing at the Teotônio cataract since about 1970, and our own data collecting there since 1976, have clearly revealed the seasonal patterns of the upstream migrations of the catfishes in the Rio Madeira, at least for the upper course of the river.

Only two families of catfishes - the Pimelodidae and Doradidae - are represented in the upstream migrations. The largest migrations are during the low water period, or between about July and November. The *dourada* (*Brachyplatystoma flavicans*, Pimelodidae), however, is known to migrate upstream between about August and March, with peaks in September and October and late December to early February. The *babão* (*Goslinia platynema*, Pimelodidae) is the only species known to migrate only during the high water period. The purpose of these upstream migrations is still unclear as the catfishes appear to pass the Rio Madeira rapids and enter eastern Bolivian rivers. Spawning is suggested, but the fishes will have to be studied in Bolivia before this can be established.

The food chain sustaining the commercial fishes

The feeding behavior of most of the important food fishes of the Rio Madeira has been studied in some detail, and it has become clear that the flooded forests play a key role in nourishing the commercial ichthyofauna (Goulding 1980). Of the nine genera that represented about 87 percent of the total commercial catch between 1977 and 1979 (see Table 4.1), at least four (*Brycon*, *Mylossoma*, *Triportheus*, and *Colossoma*) are directly dependent on the flooded forests for their foods (mostly fruits, seeds, and arthropods that fall out of the trees and into the water during the floods). These four genera represented about 36 percent of the total commercial catch during the three year period. The detritus feeding fishes (*Prochilodus*, *Semaprochilodus*, and *Curimata*) accounted for about 26 percent of the total catch, and they also appear to be heavily dependent on the flooded forests for their food in the Rio Madeira region. This statement is not based on stomach content analyses, but on the observation that after spawning at the beginning of the floods, these fishes have little body fat, but after entering the flooded forests subsequent to breeding, they build up large fat reserves that they gradually lose during the low water season when their food

(detritus) is evidently scarce. Only two piscivorous species are among the ten most important fish taxa in the commercial fisheries of the Rio Madeira. The *dourada* (*Brachyplatystoma flavicans*, Pimelodidae), however, was the single most important species between 1977 and 1979, and accounted for about 21 percent of the total catch. I have been studying the feeding behavior of this large catfish for the past three years, and the preliminary evidence - based on the examination of about 5,000 specimens - indicates that it feeds heavily on the migratory characins that, as a group, appear to be the most important fishes in the siluroid's diet. This appears to be a clearcut example of energy transference from flooded forest to migratory characin to river channel catfish. The other piscivore in the top-ten taxa is the *tucunaré* (*Cichla ocellaris*, Cichlidae), and since it is captured on the floodplain of the Rio Madeira, the food chain leading to it may begin in plankton communities, aquatic macrophytes, and flooded forests. All together, I estimate that the food chain leading to at least 75 percent of the total commercial fish catch of the Rio Madeira, as represented by the species in the 1977-1979 period, begins in the flooded forests.

The distribution of the flooded forests where the commercial fishes feed is also important. As already discussed, the Rio Madeira itself has a relatively limited and high floodplain that is inundated for a shorter time each year than are the floodplains of the lower reaches of its tributaries. Although the total floodplain area of all the tributaries taken together may be less than that of the Rio Madeira, they still offer better feeding conditions because they are inundated for a longer time. The fisheries data clearly show that most of the commercial fishes in the upper and middle Rio Madeira are in the large rightbank tributaries during the floods. This is where they find their food.

The food fishes

The two great groups of Amazon fishes are the characins (or characoids) and the catfishes (or siluroids), and together, with each about equally represented, they account for someting around 85 percent of the described species of the region (Roberts 1973). They are also the two most important groups of food fishes in the Amazon. The characins are now divided into twelve to fourteen families, depending on the classification one follows, while the catfishes embrace at least twelve families. Of the fourteen characin families recognized by Greenwood et al. (1967), only six (Characidae, Prochilodontidae, Curimatidae, Hemiodontidae, Erythrinidae and Anostomidae) are of commercial importance; the remaining families consist mostly of very small fishes. Three catfish families (Pimelodidae, Doradidae, and Hypophthalmidae) are of importance in the commercial fisheries; in the Rio Madeira region the pimelodids account for over 99 percent of the commercial catch of siluroids. The electric fishes (gymnotoids) are the third most abundant piscine group in the Amazon, accounting for about 18 percent of the described species, but they have no importance in the fisheries because most are small and stay hidden away, and there is also a cultural taboo against eating them. The large electric eel (*Electrophorous electricus*, Electrophoridae), that reaches over 2 m in length and 15 kg in weight, could be exploited to some extent, but I have found very few people in the Amazon who are willing to eat it. Cichlids acount for around two percent of the described species, and *Cichla ocellaris* is the most important food fish of this group in the Rio Madeira region, and most of the Amazon as well. There are two genera and three species of bony-tongues (Osteoglossidae) in the Amazon, of which *Arapaima gigas* and *Osteoglossum bicirrhosum* are commercial food fishes; the first has been over-exploited, while the second is mostly important in subsistence fisheries. The Clupeidae and Sciaenidae, which are mostly marine families, are represented in the Rio Madeira commercial fisheries, but they are only of modest importance to the total catch of the region. The food fish species will now be discussed in detail.

Pimelodidae

Dourada (*Brachyplatystoma flavicans* Castelnau, Pimelodidae)

The *dourada*, or gilded catfish, is named for its golden body and silvery head. Other than it distinctive color, *Brachyplatystoma flavicans* is also noted for its short barbels, which are the smallest of any of the species in its genus (Fig. 5.2). The *dourada* is a piscivorous fish, and extensive stomach content analyses and observation indicate that it feeds throughout the water column (unpublished data). Unlike most of the other large catfishes, however, *dourada* do not appear to enter flooded forests but are confined to river channels and occasionally enter floodplain lakes.

In the Rio Madeira, the *dourada* is undoubtedly the most abundant of the large siluroids, and in the 1970's it accounted for about two-thirds of the total catfish catch. It was also the single most important species - characins included - in the Rio Madeira fisheries above the mouth of the Rio Aripuanã. Other than its large biomass, there are additional factors that account for the importance of *dourada* in the commercial fisheries, namely, it is highly vulnerable to drifting deepwater-gillnets along the entire course of the river, and because it migrates upstream during both the low and high water periods, it can be exploited during a longer period of time each year than the other catfish species.

Over 90 percent of the Rio Madeira *dourada* catch is made up of fishes measuring 70-95 cm in fork length (Fig. 5.1) and weighing 4-10 kg. The evidence to date indicates that these size classes are immature fish, as no individual under about 1.2 m in fork length has been found with developed gonads. Fishermen believe that *dourada* schools originiate in or near the Rio Amazonas, and migrate up the Rio Madeira destined to the Rio Beni or Rio Mamoré in eastern Bolivia. The flooded areas of eastern Bolivia are thought by fishermen to be the spawning grounds for the *dourada*, but this region has not been investigated ichthyologically to any significant extent, and thus we do not know the nature of the fish populations there. It is worth pointing out here, however, that young *dourada* under 60 cm in fork length are very rarely caught in the Rio Madeira, and our gillnet and seine surveys of the floodplain waterbodies (including those of the tributaries) show that they are not there either. The whereabouts of the young of the *dourada*, then,

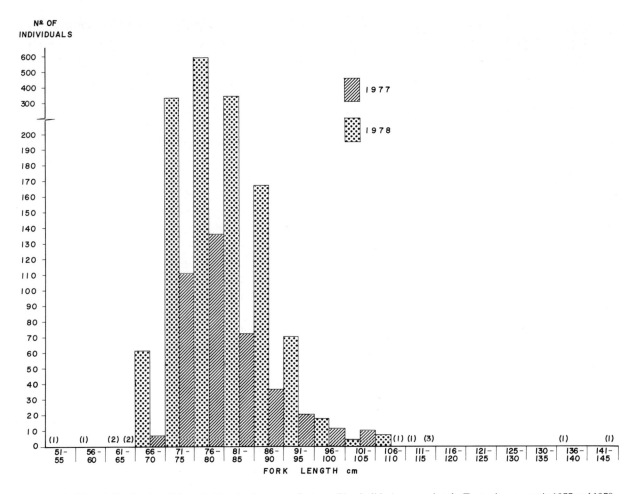

Fig. 5.1 Fork length distribution of *dourada* (*Brachyplatystoma flavicans*, Pimelodidae) captured at the Teotonio cataract in 1977 and 1978.

remain a mystery, but the eastern Bolivian flooded areas are highly suspect. There may be breeding populations along the Rio Amazonas as well, but fishermen from that area have not observed them and do not know where the young might be.

Piraíba and *Filhote* (*Brachyplatystoma filamentosum* Lichenstein, Pimelodidae)

The vernacular names *piraíba* and *filhote* may embrace more than one species, though they are taken putatively to be *Brachyplatystoma filamentosum* (Fig. 5.3 and see Fig. 3.13). The *piraíba* is the largest catfish in the Amazon basin, and rivaled only by the giant *pirarucu* (*Arapaima gigas*, Osteoglossidae) for being the largest of any freshwater species in South America. There are very few reliable measurements of large *piraíba*, and Gudger's (1943) review of mention of the fish in the literature, though excellent from an historical point of view, does not convincingly establish the size limits of the large siluroid. The medical doctor of the Roosevelt-Rondon Expedition to the Amazon reported that he saw a large *piraíba* in the town of Itacoatiara, which is on the left bank of the Rio Amazonas near the mouth of the Rio Madeira, that measured over nine feet in length (Roosevelt 1914). Roosevelt (1914) himself saw a large *piraíba* that contained the remains of a monkey in its stomach, and went on to state, "... our Brazilian friends told us that in the lower Madeira and part of the Amazon near the mouth, there is still a more gigantic catfish which in similar fashion occasionally makes prey of man' (pp. 311-312). The largest

Fig. 5.2 The *dourada* (*Brachyplatystoma flavicans*, Pimelodidae).

Fig. 5.3 The *filhote* (*Brachyplatystoma filamentosum*, Pimelodidae).

individuals that I have examined from the Rio Madeira were 1.8 and 2.1 m in fork length, and both of these individuals weighed 110 kg.

Most of the Rio Madeira catch of *Brachyplatystoma filamentosum* (sensu latu) is represented by fishes measuring 75-100 cm fork length and 5-15 kg in weight, and these are called *filhotes* by fishermen. These appear to be immature fish, as the smallest mature individual yet found was 1.2 m in fork length and 45 kg. Most of the *filhote* catch is taken with drifting deepwater-gillnets in the river channel or with castnets at the Teotônio cataract. Large *piraíba* are captured with trotlines strung to the middle of the deep channel. Following *dourada* (*Brachyplatystoma flavicans*), the *piraíba/filhote* is the most important catfish species in the Rio Madeira fisheries.

The *piraíba/filhote* is considered a second class food fish in the Rio Madeira region and, like the other large catfishes, most of it was exported to Southern and Southeastern Brazil in the 1970's. The flesh of *piraíba* especially, and *filhote* less so, is said to be *remoso*, or imbued with pathogenic properties that aggravate inflammation and cause other illnesses, though there is no evidence for this. It is also widely believed in the Amazon that the liver of the *piraíba* is toxic, and that not even vultures will touch it. In 1977, in the village of Calama on the right bank of the rio Madeira, I was called one day to treat identical illnesses of two fishermen employed by a Porto Velho refrigeration company. When I arrived on the fishing raft where they were stationed, I found both of them suffering miserably from high fever and serious skin inflammation over their entire bodies. They appeared to be burned, but upon questioning, I found that they had both come down with the same symptoms a few hours after eating the liver of *piraíba*, a dish which even their fellow fishermen had refused to partake in and which they warned their novice colleagues not to eat. In the two or three days that followed, the two sufferers shed the upper layer of their skin and one, who ventured into the sun, returned in extreme pain. Within about two weeks, both of the ailing fishermen had recovered. The evidence and symptoms suggest vitamin A poisoning, and the *piraíba* liver should be studied biochemically.

Piramutaba (*Brachyplatystoma vaillantii* Valenciennes, Pimelodidae)

Brachyplatystoma vaillantii is very similar in appearance to *Brachyplatystoma filamentosum*, but is easily distinguished from it by its much larger adipose fin (Fig. 5.4). It is also a smaller catfish; the largest individuals that I examined from the Rio Madeira were no more than about 80 cm in fork length and 10 kg in weight. According to residents and fishermen at the Teotônio cataracts, schools of *piramutaba* appear at the Rio Madeira rapids every four or five years. Refrigeration company records indicate that the catfish was abundant at the Teotônio cataract in 1974 (as does also a popular postcard circulated in Porto Velho that shows large quantities of it being salted on the rocks near the rapids), and our data show that schools of it appeared again in 1979 during the low water season (October and early November). About 30,000 tons of *piramutaba* are taken near the Amazon estuary each year, and this makes it the single most important species in all of the Amazon basin. Unfortunately the species or the fisheries for it have never been studied in detail, but it is known that schools originating in the estuary migrate upstream in the Rio Amazonas and indeed well up the Rio Solimões as well; commercial fishermen accompany these schools and exploit them. The purpose of these large upstream migrations is unknown, as is the whereabouts of the nursery habitats for the young of *piramutaba*. The *piramutaba* schools that arrive at the Rio Madeira rapids every four or five years may begin their migrations from the Amazon estuary, though they may not take that long to make the 1,500 km journey.

Babão (*Goslinia platynema* Boulenger, Pimelodidae)

Babão may be translated from the Portuguese as 'slobberer', and *Goslinia platynema* apparently carries this vernacular appellation because fishermen think its elongated, wide, and flattened mental barbels look like mucus hanging from the fish's mouth (Fig. 5.5). The species reaches at least 80 cm in fork length and 5 kg in weight. It is not a very colorful catfish, being grey to black dorsally and

Fig. 5.4 Piramutaba (*Brachyplatystoma vaillantii*, Pimelodidae). About 55 cm fork length.

Fig. 5.5 The *babão* (*Goslinia platynema*, Pimelodidae). About 75 cm fork length.

dirty white ventrally. Its large barbels appear to make up for its extremely small eyes, but nothing is known about its feeding behavior.

The only known fishery for *Goslinia platynema* in the Amazon basin is at the Teotônio cataract of the Rio Madeira. The species is poorly known, even to fishermen and riparian folk, elsewhere in the Amazon. The *babão* appears in schools at the Teotônio cataract during the high water period, with most being caught between February and April. The catfish is captured with gaffs when it tries to work its way upstream along the left bank, or with drifting deepwater-gillnets in the large bay area below the rapids. Most of the *babão* catch at Teotônio consists of adult fish, many of which are ripe, and this suggests that the upstream moving schools may be destined to spawning grounds in the Rio Beni or Rio Mamoré swampy areas in eastern Bolivia. During the low water period the species is only occasionally captured at the Rio Madeira rapids and in the rest of the river below them. A little over nine tons of *Goslinia platynema* were captured at the Teotônio cataract in 1977.

Bico de Pato (*Sorubim lima* Schneider, Pimelodidac)

The common name *bico de pato*, or duck's beak, is an allusion to the greatly elongated snout, inferior mouth with projecting upper jaw, and widely separated, laterally displaced eyes of *Sorubim lima* (Fig. 5.6). The catfish reaches at least 40 cm in fork length. Nothing much is know about the feeding behavior of *Sorubim lima*, a widely distributed fish in South America, though Ringuelet et al. (1967) report that in Argentinian waters it feeds on small fishes and crustaceans. Large schools of *Sorubim lima* are occasionally encountered in the Rio Madeira, but the market value of the species does not make it very attractive to fishermen. Schools of *Sorubim lima* appear at the Teotônio rapids towards the end of the low water season and at the beginning of the floods. Cataract fishermen capture modest amounts of them with seines when they become concentrated in the pools below the rapids.

Mandi (*Pimelodus* spp. and *Pimelodella* spp., Pimelodidae)

The taxonomy of these small catfishes (about 15-35 cm fork length) is poorly understood, and for the present most forms can only be discussed at the generic level. *Pimelodus* and *Pimelodella* fishes are common along riverbanks but almost nothing is known of their feeding behavior in Amazonian waters. *Pimelodus blochii*, a species that I studied in the Rio Machado, feeds on fruits, insects, and detritus, and others may be equally omnivorous (Fig. 5.7). This is also the only species that I have encountered in schools in the Rio Madeira, but there must be others as relatively large quantities of several taxa appear in the Porto Velho market occasionally. Fishermen are loath to capture small pimelodid catfishes because most of them have hardened dorsal and pectoral fin spines that get stuck in the mesh of seines. A captured school of *mandi* may take several hours to remove, and usually with considerable harm to the net. In general these taxa are probably underexploited in the fisheries, and until a better method is found for dealing with them, they will probably remain so.

Jaú (*Paulicea lutkeni* Steindachner, Pimelodidae)

Paulicea lutkeni is one of the largest South American freshwater catfishes, and is found from the La Plata to the Amazon. In girth it is the grossest neotropical catfish, and has a short, thick body with a very large head (Fig. 5.8). Ihering (1929) reported that the largest *jaú* in São Paulo state reached no more than about 1.5 m in length, but could weigh up to 150 kg. I have not seen any *jaú* this large in the Amazon, though fishermen report that they are occasionally caught. The largest individual that I examined from the Rio Madeira was no more than about 1.3 m in standard length and 100 kg. The large catfish appears to be mostly a piscivore, though fishermen report that smaller individuals that enter flooded forests also feed on fleshy fruits. In color the Rio Madeira *jaú* is a drab olive-green when adult; smaller individuals have light spots scattered dorsally across their bodies. The general unattractiveness of the *jaú*, and its poor

Fig. 5.6 The *bico de pato* (*Sorubim lima*, Pimelodidae). About 30 cm fork length.

Fig. 5.7 The *mandi* (*Pimelodus blochii*, Pimelodidae). About 18 cm fork length.

Fig. 5.8 A *jaú* (*Paulicea lutkeni*, Pimelodidae) captured at the Teotônio cataract.

standing in local folklore, have given it perhaps the lowest reputation of any fish in the Rio Madeira region. The large catfish is reported to cause all types of skin diseases, inflammations, miscarriages, and hemorrhoids if eaten. Refrigeration companies, however, found a market for it in the early 1970's. In 1974, at least 25 tons of *jaú* were captured at the Teotônio cataract, but since that time yields have declined to below 10 tons.

Jaú begin appearing at the Teotônio cataracts in July or August, usually a month or so before the other large pimelodid catfishes arrive there. Its behavior at the rapids is also different than the other large pimelodids. *Brachyplatystoma* and *Pseudoplatystoma* do not appear to feed very much when migrating through the Rio Madeira rapids, as their stomachs are usually empty. The *jaú*, however, appears to migrate to the rapids during the low water period to feed on upstream moving characins that become concentrated in the pools below the rocks that are being swept across by a strong current. The large catfish can easily be seen pursuing the characins in the turbulent water. As discussed earlier, fishermen bait the hooks of their handlines with the characins that are trying to get through the rapids and on which the *jaú* is feeding. *Prochilodus nigricans* is the most common bait used, and it is interesting to note that Ihering (1929) also

Fig. 5.9 The *caparari (Pseudoplatystoma tigrinum*, Pimelodidae).

mentioned that a very similar species (*Prochilodus scrofa*) was often seen being pursued by *Paulicea lutkeni* in São Paulo rivers.

From studying refrigeration company receipts in Porto Velho, I estimated that about 20-30 tons of *jaú* were captured annually at the Teotônio cataract between 1971 and 1974. Most of the large *jaú*, according to fishermen and fishmongers, were captured in the first years of heavy exploitation. Unfortunately we have no measurements for these large fishes. The average *jaú* caught in 1977 at the Teotônio cataract measured about 1.0-1.3 m in fork length and weighed 30-45 kg.

Caparari (Pseudoplatystoma tigrinum, Valenciennes, Pimelodidae)

Pseudoplatystoma tigrinum is a tiger-striped catfish that reaches at least 1.3 m in standard length and 20 kg in weight (Fig. 5.9). It has an elongated, dorsally-ventrally flattened head, and relatively short barbels. The species is abundant in many types of habitats, including flooded forests, river channels, lakes, and floating meadows. It migrates upstream in the Rio Madeira during the low water period and is captured in the main channel with drifting deepwater-gillnets and at the Teotônio cataracts with castnets when it becomes concen-

Fig. 5.10 The *surubim* (*Pseudoplatystoma fasciatum*, Pimelodidae).

trated in the pools below the crest of the rapids. Catches of *Pseudoplatystoma tigrinum* have declined considerably since the large-scale exploitation of catfishes began in the early 1970's. The species appears to be highly vulnerable to drifting deepwater-gillnets, and most of the larger individuals – which undoubtedly accounted for a major part of the catches in the initial years – have probably already been caught. In the Rio Mamoré and Rio Guaporé, the species began to be heavily exploited in the late 1970's, and in 1978 it was the most important catfish that was exported from this new fisheries frontier.

Surubim or *Sorubim* (*Pseudoplatystoma fasciatum* Linnaeus, Pimelodidae)

Pseudoplatystoma fasciatum is very similar to its congener, *P. tigrinum*, but can be easily distinguished from it by its thinner, usually vertical stripes, and narrower snout (Fig. 5.10). *Pseudoplatystoma fasciatum* is also smaller and the largest individuals that I have measured from the Rio Madeira were about 90 cm in fork length. There are no apparent ecological differences between *P. fasciatum* and *tigrinum*, and the two species are often caught together when they are migrating upstream in the Rio Madeira during the low water period.

Fig. 5.11 The *coroatá* (*Platynematichthys notatus*, Pimelodidae).

Like *P. tigrinum*, *P. fasciatum* catches have declined considerably since large scale commercial exploitation began.

Coroatá, Coronel, or *Cara de Gato* (*Platynematichthys notatus* Schomburgk, Pimelodidae)

Platynematichthys notatus travels under three vernacular names, of which *coroatá* is said to be the most correct, while *coronel* is a military nickname which need not be discussed here, and *cara de gato*, or cat'a face, an allustion to the feline appearance of its broad and bewhiskered snout .Fig. 5.11). It is dorsally and anteriorly spotted and has a large black patch on the inferior lobe of its caudal fin. It attains fork lengths of at least 80 cm. The species is widespread and found in many types of habitats in both the Rio Madeira and its clearwater tributaries. It is often caught with hook-and-line by subsistence fishermen, while commercial catches are usually taken with seines when the catfishes come mixed with schools of characins on which they probably are feeding.

Pirarara (*Phractocephalus hemiliopterus* Spix, Pimelodidae)

Phractocephalus hemiliopterus is a colorful catfish

characterized by its strongly ossified head, huge and bony pre-dorsal plate, marked countershading, and bright orange tipped caudal, adipose, dorsal, and anal fins (Fig. 5.12). It reaches at least 1.3 m in fork length and 80 kg in weight. It is known to feed on fish, fruits, and crabs. The species was relatively more important in the initial years of large-scale siluroid exploitation in the Rio Madeira, but most of the old and large *pirarara* have already been caught and the individuals captured now are usually under 80 cm and 10 kg. The species is highly vulnerable to trotlines, and this is the gear with which most of them are captured in the Rio Madeira fisheries.

Dourada Fita (*Merodontotus tigrinus* Britski, Pimelodidae)

Merodontotus tigrinus was unknown to science until Dr. Heraldo Britski of the Museum of Zoology of the University of São Paulo discovered it in my Rio Madeira fish collection in 1978 (Fig. 5.15). It is the only large catfish that was not described in the nineteenth century. Superficially *Merodontotus tigrinus* looks like *Brachyplatystoma juruense*, because it has a yellow body with black stripes. The black stripes of *Merodontotus tigrinus*, however, are continuous when viewed from the profile, whereas some of those of *Brachyplatystoma juruense* are broken or divided. Its dentition is also quite different and hence the erection of a new genus for it (Britski, in press). The species appears at the Teotônio cataract during the low and rising water period, and fishermen believe it comes mixed with *Brachyplatystoma flavicans*, though it would be hard to see how it mimics it with its distinctively different color. The species apparently has a small biomass and is not very important as a food fish, but is worth mentioning here because of its size and recent discovery.

Peixe Lenha (*Surubimichthys planiceps* Agassiz, Pimelodidae)

Peixe lenha, or the firewood catfish, is not an exciting name for *Surubimichthys planiceps*, which is an attractive siluroid characterized by its very elongated and flattened head, thin cylindrical body, black spots on its dorsal and ventral regions which are separated by a wide white patch that runs from the opercular opening to the medium caudal fin rays, and long maxillary barbels (Fig. 5.14). *Surubimichthys planiceps* reaches at least one meter in fork length, but its thin body and large head do not leave much to eat once the fish has been eviscerated and decapitated. The species is captured mostly during the low water period at the Teotônio cataract when it is migrating upstream in the Rio Madeira.

Dourada Zebra (*Brachyplatystoma juruense* Boulenger, Pimelodidae)

Brachyplatystoma juruense is highly worthy of its vernacular name, *dourada zebra*, or gilded zebra (Fig. 5.13). Its yellow body is adorned with thick black stripes. The attractive catfish reaches at least 60 m in fork length. It is unclear whether *Brachyplatystoma juruense* migrates upstream in the Rio Madeira in schools with its own species or mixed with *Brachyplatystoma flavicans*. In any case, it is captured in small quantities during the low water season and beginning of the floods at the Teotônio cataract. Fisheries data indicate that it does not have a very large biomass.

Barba-Chata (*Pinirampus pirinampu* Spix, Pimelodidae)

The vernacular name, *barba-chata*, or flat-whisker, succinctly indicates one of the distinguishing characters of *Pinirampus pirinampu*, that is, its greatly flattened barbels (Fig. 5.16). It is a husky catfish and reaches at least 60 cm in fork length. Other than its long and flattened barbels, it is also noted for its extremely long adipose fin. It is not a very colorful catfish, though when first removed from the water there is sometimes a bluish-green sheen dorsally. It is common along riverbanks and village and city waterfronts where, like *Callophysus macropterus*, it scavenges what it can find and is often caught by children and subsistence fishermen. It appears in schools at the Teotônio cataract in November and December, and is there captured with seines.

Fig. 5.12 The pirarara (*Phractocephalus hemiliopterus*, Pimelodidae). Photo by Barbara Gibbs.

Fig. 5.13 The *dourada zebra* (*Brachyplatystoma juruense*, Pimelodidae).

Fig. 5.14 The *peixe lenha* (*Surubimichthys planiceps*, Pimelodidae).

Fig. 5.15 *Dourada fita* (*Merodontotus tigrinus*, Pimelodidae). Photograph courtesy of Dr. Heraldo Britski.

Fig. 5.16 The *barba-chata* (*Pinirampus pirinampu*, Pimelodidae).

Because of its abundance along riverbanks, it is more common in subsistence fisheries.

Pintadinho or *Piracatinga* (*Callophysus macropterus* Lichenstein, Pimelodidae)

Callophysus macropterus is a medium sized catfish reaching at least 50 cm in fork length, and is characterized by its (usually) spotted body, long adipose fin, long barbels, and incisive to semi-cusped dentition (Fig. 5.17). It is one of the most common species along riverbanks (both in the Rio Madeira and its clearwater tributaries) and is usually found in relatively large concentrations near village and city waterfronts where it plays the role of aquatic vulture, scavenging up offal, human excrement, and the like. Relatively large schools of *Callophysus macropterus* appear at the Teotônio cataract between about October and mid-December, and are sometimes captured with seines and sold in Porto Velho.

Other than its modest role as a food fish, *Callophysus macropterus* has a second billing in the Rio Madeira fisheries. The species can be a voracious piscivore (but it also eats fruits in the flooded forests) and commonly attacks fishermens' catches in seines or nets, or catfishes gaffed or hooked on trotlines. Often in union with similar acting trichomycterid and cetopsid catfishes, it rips out small pieces of the fishes that it attacks (see Chapter 3). These fishes are so aggressive, or hungry, that they often refuse to let go of the victim even after it has

Fig. 5.17 The *pintadinho* or *piracatinga* (*Callophysus macropterus*, Pimelodidae). About 20 cm standard length.

been removed from the water, all the time tearing away at its flesh with their sharp dentition. The broad feeding spectrum of *Callophysus macropterus*, and its aggressive feeding behavior, suggest that it might be a good choice for experimental fish culture.

Doradidae

Cuiu-Cuiu (Oxydoras niger Kner, Doradidae)

Oxydoras niger appears to be the largest doradid, and reaches at least 1.2 m in fork length and 20 kg in weight (Fig. 5.18). It is a toothless form with a downturned mouth that it uses for muzzling into detritus. Stomach contents consist of detritus (especially leafy material in decomposition), which is usually invested with insect larvae such as midges (*Chironomus* spp., Chironomidae - Diptera) and mayflies (Ephermeroptera), and shrimps. It inhabits a wide variety of habitats including flooded forest, floodplain lakes, and is sometimes caught while in schools and moving upstream in the Rio Madeira. Small schools appear at the Teotônio rapids in June and July and, according to fishermen, in some years even further into the low water season. The *cuiu-cuiu* is appreciated as a food fish in Porto Velho but is only of very modest importance in the commercial catches.

Bacu Comum (Pterodoras granulosus Valenciennes, Doradidae)

Pterodoras granulosus reaches at least 60 cm fork length and is brownish-tan in color (Fig. 5.19). It appears to be restricted mostly to the Rio Madeira and its floodplain, and it was never caught in our surveys of the clearwater tributaries. On the Rio Madeira floodplain, I have caught individuals that were stuffed full of fleshy fruits (especially *Astrocaryum jauary*, Palmae). The species also feeds on snails and aquatic macrophytes. Unlike the other large doradids that appear at the Rio Madeira rapids at the end of the floods, *Pterodoras granulosus* shows up, and in relatively larger schools, at the beginning of the annual inundation in November and December and occasionally in early January. Cataract fishermen capture modest quantities of them with seines when the fishes become concentrated in the pools below the crest of the Teotônio rapids. *Pterodoras granulosus* is welcomed in Porto Velho as a food fish and finds a ready market.

Fig. 5.18 The *cuiu-cuiu* (*Oxydoras niger*, Doradidae).

Bacu or *Bacu Rebeca* (*Megaladoras irwini* Eigenmann, Doradidae)

Megaladoras irwini is one of the largest doradids, reaching at least 60 cm in fork length, and it is heavily armored with lateral scutes and greatly thickened dorsal and pectoral fin spines (Fig. 5.20). Its ground color is brown, and its body is often adorned with black spots, especially on the stomach. It is known to feed heavily on fruits (e.g. *Licania longitpetala*, Chrysobalanaceae and *Astrocaryum jauary*, Palmae) and pulmonate snails in the flooded forests during the high water season. The large doradid is a good seed dispersal agent as it swallows fleshy fruits whole; the seeds pass through the intestinal system unharmed. Small schools of *Megaladoras irwini* appear at the Teotônio cataract above Porto Velho in June or July and are sometimes caught by commercial fishermen, though their market value is rather low. Drifting deep-water-gillnet fishermen often catch *cuiu-cuiu* in the Rio Madeira during the low water period, but they usually turn them loose because of their low market value.

Bacu Pedra (*Lithodoras dorsalis* Valenciennes, Doradidae)

Bacu pedra or rock-*bacu*, is an appropriate vernacular name for *Lithodoras dorsalis*, which is the most heavily armored doradid species (Fig. 5.21). Thick bony plates cover almost the entire body of

Fig. 5.19 The *bacu comum (Pterodoras granulosus*, Doradidae). The greatly distended stomach of the fish shown above is stuffed full of palm fruits *(Astrocaryum jauary)*, samples of which are shown in the bowl to the left.

Fig. 5.20 The *bacu* or *bacu rebeca (Megaladoras irwini*, Doradidae).

this large fish, including the ventral area which is mostly naked in other doradid fishes. *Lithodoras dorsalis* reaches at least 90 cm fork length and 12 kg in weight. The species is known to feed on fruits (e.g. *Licania longipetala*, Chrysobalanaceae) in the flooded forests and on grass blades (*Paspalum repens*, Gramineae) of macrophytes that grow along the levee of the Rio Madeira. Like *Megaladoras irwini*, it is also a seed dispersal agent of the fleshy fruits that it eats. Small schools of *Lithodoras dorsalis* appear at the Teotônio cataract at the same time as those of *Megaladoras irwini* in June and July, and likewise, are occasionally taken by commercial fishermen. When caught in drifting deep-water-gillnets they are usually killed and disposed of because refrigeration companies refuse to buy them.

Loricariidae

Bodó (*Plecostomus* and *Pterygoplichthys*, Loricariidae)

The loricariids are armored catfishes with inferior mouths which they use to muzzle into detritus or remove fine particles from substrates such as submerged tree trunks. Little is known in detail about their feeding behavior, though *Plecostomus* (Fig. 5.22) and *Pterygoplichthys*, the most important food fish genera of the Rio Madeira region, often eat detritus invested with insect larvae, such as bloodworms (*Chironomus* spp., Chironomidae - Diptera). The species of *Plecostomus* and *Pterygoplichthys* exploited in the Rio Madeira floodplain average about 25-40 cm in fork length. These fishes are captured with castnets during the low water period when the floodplain lakes and lagoons shrink and the fishes become concentrated. Both of the loricariid genera are air-breathers, and are easily spotted by fishermen when they surface. Subsistence fishermen often catch large numbers of *bodó* during the low water period and then store them in canoes half-filled with water or in oil drums kept at the side of their houses. The hardy fishes are able to live several weeks under such conditions, and thus a living protein supply can be stored for several weeks.

Hypophthalmidae

Mapará (*Hypophthalmus edentatus* Spix and *perporosus* Cope, Hypophthalmidae)

Hypophthalmus edentatus (Fig. 5.23) and *perporosus* are naked catfishes most noted for their long anal fins and numerous, fine gillrakers. *Hypophthalmus edentatus* has been shown to feed on crustacean zooplankton (Carvalho 1979), and *H. perporosus* is thought to be planktivorous as well. In the Rio Madeira region, *mapará* are captured mostly in June, July, and August, especially in the mouth area of the Rio Jamari, a clearwater affluent near Porto Velho. There appears to be some plankton production in the lower reaches of the Rio Jamari, when water level is dropping, and the *mapará* may move into this area to feed. Schools are also occasionally caught migrating upstream in the Rio Madeira.

Characidae

Jatuarana (*Brycon* sp., Characidae)

The taxonomy of the genus *Brycon* is in much confusion and no reliable scientific specific names can be given in the present state of knowledge. The species called *jatuarana* by Rio Madeira fishermen is a silvery, irridescent fish with a pink sheen, and it often has a black band running obliquely from the caudal to the end of the anal fin (Fig. 5.24). In general appearance *jatuarana* somewhat resemble trout, though they are far removed from them in phylogeny. *Jatuarana* of the Rio Madeira valley reach at least 65 cm in standard length and 4 kg in weight. The species has a complex set of dentition consisting of peg-like teeth in the first row of the pre-maxillae, and behind these are two rows of multicusped teeth; in the lower jaw there are broad-based multicusped teeth with two smaller, conical teeth behind the front symphyseal pair. The maxillae are also fully toothed. The *jatuarana* is an omnivorous fish with a strong preference for fruits and seeds which it procures in the flooded forests during the annual inundation.

The *jatuarana* was the single most important

Fig. 5.21 The *bacu pedra* (*Lithodoras dorsalis*, Doradidae).

food fish of the characin group in the Rio Madeira fisheries in the late 1970's, and fishermen and fishmongers reported that it has held that position since large-scale commercial operations began in 1959. In the period 1977-1979, however, the *dourada* catfish (*Brachyplatystoma flavicans*, Pimelodidae) was the single most important species, when all taxa are considered. Subsequent to the annual flood, populations of *jatuarana* first begin migrating in late May or early June, and these appear to be fishes abandoning the small affluents of the Rio Madeira; since the flooded forests of these smaller tributaries are drained faster than those of the larger affluents, and their channels are smaller and more restricted, the migratory fishes may in fact be fleeing to their own safety. The main migrations down the larger clearwater tributaries are in July and August, and these fishes are heavily exploited in the affluent mouths or when migrating upstream in the Rio Madeira (Fig. 5.25). By October, most of the *jatuarana* schools have entered another tributary farther upstream and catches are minimal until the spawning runs in late December and early January. The exploitation of *jatuarana* during their spawning migrations was discussed in detail in Chapter 3.

The *jatuarana* is considered a first-class food fish, and much effort is devoted to catching it. Despite

Fig. 5.22 The *bodó* (*Plecostomus* sp., Loricariidae).

this effort, it has held up fairly well in comparison to the other large food fish characins, namely, *Colossoma macropomum* and *Colossoma bidens*.

Matrinchão (*Brycon* sp., Characidae)

The fish called *matrinchão* by Rio Madeira fishermen is a member of the genus *Brycon*, and the species reaches only about 30 cm in standard length (Fig. 5.26). It is most commonly found in the clearwater tributaries where during the floods it can be seen feeding on fruits, seeds, and arthropods that fall into the water. During low water, it takes refuge in the river channel. The *matrinchão* appears to spawn in the clearwater tributaries as they are not seen descending to the Rio Madeira, as do the migratory characins. The young of *matrinchão* are also observed in the clearwater tributaries, suggesting also that adults spawn here. *Matrinchão* are captured mostly during the low water period when they are found mixed with schools of their congener called *jatuarana*, the latter always in much greater abundance. It appears, then, that they disperse from one tributary to another farther upstream as mimics of the *jatuarana*.

Fig. 5.23 The *mapará (Hypophthalmus edentatus*, Hypophthalmidae). About 25 cm fork length.

Fig. 5.24 The *jatuarana* (*Brycon* sp., Characidae). About 34 cm standard length.

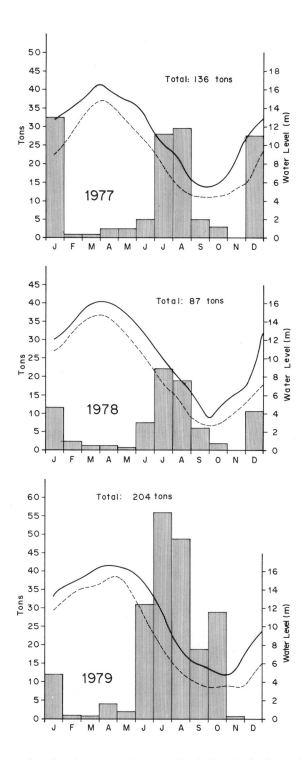

Fig. 5.25 Monthly distribution of annual catches of *jatuarana (Brycon* sp.) by the Porto Velho fleet in the Rio Madeira in 1977, 1978, and 1979. Note that the monthly distribution of catches is similar in all three years, and this reflects clearly the migration patterns of the *jatuarana* in relation to water level. In 1979 the exploitation of spawning *jatuarana* was prohibited, and thus there were no catches of the species in December. It should be noted, however, that this had very little effect on the total anual catch in comparison to 1977 and 1978. The solid and broken lines represent, respectively, the monthly high and low water levels.

Fig. 5.26 The *matrinchão* (*Brycon* sp., Characidae). About 23 cm standard length.

Fig. 5.27 The *tambaqui* (*Colossoma macropomum*, Characidae).

Tambaqui (*Colossoma macropomum* Cuvier, Characidae)

The *tambaqui* is the largest scaled fish in the Amazon basin, reaching at least 90 cm in standard length and 30 kg in weight (Fig. 5.27). It is a deep-bodied fish and often displays distinctive counter-shading, being usually black ventrally and golden to olive or moss-green dorsally. It possesses large molariform-like, multicusped teeth that can be used to crack and masticate hard nuts, such as those of rubber tree fruits (*Hevea spurceana* and *H. brasiliensis*, Euphorbiaceae) and palm fruits (*Astrocaryum jauary*, Palmae). Another outstanding feature of the *tambaqui* is the numerous and fine gillrakers, and these are employed, especially in young fish, to capture zooplankton; for older individuals fruits and seeds are the main component of the diet, and zooplankton is only of minimal importance.

In the Rio Madeira valley, *tambaqui* migrate out of the clearwater tributaries at the end of the floods and enter the principal river. In the clearwater affluents they apparently travel in deeper water than most migratory characins, as they are only rarely seen by fishermen in the tributary mouths, but first spotted when they enter the woody shore areas along the Rio Madeira. During the low water period, *tambaqui* schools take refuge in the woody shore areas, or *pauzadas*, of the Rio Madeira. In this habitat they are difficult to catch because seines cannot be payed without becoming snagged on a piece of wood. Fishermen occasionally attempt to frighten small groups of *tambaqui* out into open water, but the large fishes appear to be tenacious of the woody areas and are not easily fooled. In the late 1970's, there was no evidence of any large schools of *tambaqui* being caught in the Rio Madeira above the mouth of the Rio Aripuanã during the *piracema* migrations of the other migratory characins (mostly between August and October). Commercial fishermen report, however, that before about 1975 relatively large captures of *tambaqui* were also made during the *piracema* migrations, but that after that year the species became even more tenacious of woody shore areas. I believe that the *tambaqui* has been over-exploited in the Rio Madeira valley, and it is not unreasonable to assume that there has been selection – through the agent of commercial fishing – for those populations that escaped capture by hiding in the woody shore areas and spending the least time possible in the open water of the river where they are highly susceptible to capture with seines.

In late October and November, when water level is rising, *tambaqui* schools begin to abandon the woody shore areas and migrate upstream in the Rio Madeira. At this time these fishes are very ripe, and the upstream migrations are undoubtedly related to spawning, though it is not known where the *tambaqui* spawns (fishermen state that it is along the grassyevees that are being inundated with rising water). It is during these upstream migrations that most *tambaqui* are captured by commercial fishermen. Though water level is rising, there are still beaches in October and early November, and fishermen attempt to accompany a *tambaqui* school moving upstream until it passes near a beach where 200-300 m seines can be used to encircle the fish and then pull them ashore. The largest catch that I have verified was about 10 tons, or 1000 individuals. After spawning, the *tambaqui* migrate to the clearwater tributaries where the floodplain forests are being inundated, and where they feed heavily for four or five months when fruits and seeds are available in abundance.

Until recently the *tambaqui* was little threatened in the flooded forests by subsistence and commercial fishermen. The diffusion of gillnets, however, has changed that. The *tambaqui* is especially vulnerable to gillnet fishing because it is a deep-bodied species that becomes easily entangled; furthermore, because of its size, large mesh gillnets are used, and these suffer little damage from *piranhas*, which are much smaller fishes and thus do not get entangled. Petrere (1978b) has shown that gillnets account for most of the commercial catch of *tambaqui* arriving in the Manaus (Amazonas) market, and indeed in the late 1970's it was the most important single species caught in the commercial fisheries of the state of Amazonas. In 1976 it represented about 44 percent of the total Manaus catch, excluding large catfishes that were exported and not considered in Petrere's (1978b) study. It appears highly unlikely that the *tambaqui* fishery will for very long survive, considering the widespread and intensive commercial and subsistence effort that is being devoted to it.

Most of the *tambaqui* consumed in Porto Velho in the late 1970's came from the Rio Mamoré in Bolivia. Bolivian fishermen were capturing these fishes with gillnets and transporting them downstream to Guajará-Mirim where they were sold to Brazilian refrigeration companies that trucked them to Porto Velho. In 1979, at least 120 tons of *tambaqui* were exported from the Rio Mamoré (mostly) and Rio Guaporé to Porto Velho, and probably that much more to Rio Branco, Acre.

Pirapitinga (*Colossoma bidens* Spix, Characidae)

Colossoma bidens is the second largest scaled fish in the Amazon basin, reaching at least 85 cm in standard length and 20 kg in weight. In color it ranges from light to steel blue (Fig. 5.28). Its teeth are similar to those of *Colossoma macropomum*, except that the first and second rows in the upper jaw are separated; the *pirapitinga* also has a few maxillary teeth, whereas the *tambaqui* has none. The *pirapitinga* and *tambaqui* also eat the same fruit and seed species, and there is much overlap in their diets when they are in the flooded forests.

The migratory behavior of *Colossoma bidens* is not as clear as that of its congener, and because the species has been over-exploited in the Rio Madeira basin, it is difficult to establish patterns of movement as evidenced by commercial exploitation of the species. Fishermen and fishmongers testify that the *pirapitinga* was one of the most important commercial species before about 1974; in the late 1970's, only a few tons of *pirapitinga* were being caught each year, and most of this was represented by fishes captured from upstream migrating schools located in the Rio Madeira near the Rio Aripuanã. Like the *tambaqui*, the *pirapitinga* is threatened by the diffusion of gillnets, and its importance as a commercial food fish in the Rio Madeira valley will likely be diminished to insignificance within the next several years.

Pacu Vermelho (*Mylossoma albiscopus* Cope), *Pacu Toba* (*Mylossoma duriventris* Cuvier), and *Pacu Branco* (*Mylossoma aureus* Agassiz, Characidae)

Pacu of the genus *Mylossoma* are deep-bodied, greatly compressed fishes with small projecting heads (Fig. 5.29). *Mylossoma duriventris* is the largest species, reaching at least 30 cm in standard length; the other two forms in the Rio Madeira fisheries are slightly smaller. In color they are silvery, irridescent, and, depending on the species and the water type, they have varying degrees of rusty to bright orange adorning their anal and caudal fins, cheeks, and humeral and ventral regions. Though they have relatively small mouths, their jaws and dentition are strongly built. They have two rows of teeth in the upper jaws, and two small conical members behind the symphyseal pair of the front row in the lower jaw. For the most part the teeth are broad-based with rather sharp-edged, projecting cusps. The feeding behavior of *Mylossoma albiscopus* and *M. duriventris* is very similar, and both species are mainly seed and fruit eaters, but also include leaves and arthropods that fall into the water in their diets. *Mylossoma aureus* appears to favor the floodplain areas of the Rio Madeira, rather than its tributaries. The feeding behavior of species has been studied in an area of the Rio Solimões' floodplain, and its was shown to feed mostly on seeds and fruits as well (Paixão 1980).

Mylossoma is one of the five most important food fish taxa in the Rio Madeira fisheries. It is difficult for data collectors to distinguish the species, but I would guess, from observing mixed *pacu* catches in the Porto Velho market, that *Mylossoma albiscopus* was the most important form captured in the period 1977-1979 in the Rio Madeira region. Though the *pacu* have only a meager offering of flesh, they are considered delicious, especially when pan-fried. They are captured mostly during the low water period when they migrate out of the tributaries and upstream in the Rio Madeira, and less so during the spawning period at the beginning of the floods.

Pacu Mafurá (*Myleus* spp., Characidae)

In the Rio Madeira basin the genus *Myleus* is confined mostly to the tributaries of the main river,

Fig. 5.28 The *pirapitinga (Colossoma bidens*, Characidae). About 65 cm standard length.

Fig. 5.29 A. The *pacu vermelho (Mylossoma albiscopus*, Characidae). About 16 cm standard length. B. The *pacu branco (Mylossoma aureus)*. About 13 cm standard length.

and appears to avoid turbid water. There are at least four species of *Myleus* in the large rightbank clearwater tributaries of the Rio Madeira, but the taxonomy of these fishes is very poorly known. *Myleus* fishes are easily distinguished from their relatives of the genus *Mylossoma* by their falcate (female) and bilobed (male) anal fins and their predorsal spine (Fig. 5.30). The largest species of *Myleus* caught in the commercial fisheries reach about 35 cm in standard length. During the high water period *Myleus* fishes are captured with pole-and-line using fruit-baited hooks (especially fruits of the riverside shrub *Amanoa* sp., Euphorbiaceae), and during the low water season with gillnets drifted across the beaches (see Chapter 3). As food fishes they are considered inferior to *Mylossoma* but nevertheless have a market in Porto Velho.

Sardinha (*Triportheus elongatus* Gunther, *Triportheus angulatus* Agassiz, and *Triportheus* spp., Characidae)

Fishes of the genus *Triportheus* are deep-keeled characins with greatly expanded pectoral fins. They have very large scales and are silvery in color. The *sardinha comprida* (*Triportheus elongatus*) is the largest species, with adults ranging between about 15-25 cm in standard length; *sardinha chata* (*Triportheus angulatus*) is smaller and deeper keeled than the other forms (Fig. 5.31). Other than these two common taxa, there are at least two additional species in the Rio Madeira basin. I have examined a few specimens of one of the species that had zooplankton in their stomachs. The fourth species appears to be the smallest, but I have no ecological information on it.

Triportheus elongatus and *agulatus*, the two most important commercial species, live near the surface and their greatly expanded pectoral anatomy endows them with admirable upward thrust which they employ to snap up fruits, seeds, and arthropods that fall into the water. They have rather weak jaws, but numerous multicusped teeth that they use for grasping and cutting. Like other fruit and seed eating fishes, they are highly dependent on the flooded forests for their food.

Sardinhas are captured mostly during the low water period when they are encountered moving out of the tributaries or upstream in the Rio Madeira. Though regionally they are considered a delicious fish, their small size reduces their market price, and hence fishermen only take them if other, more valuable, species cannot be found.

Piranha Caju (*Serrasalmus nattereri* Kner, Characidae)

Serrasalmus nattereri is a robust *piranha* noted for its blunt snout, large and extremely sharp-teeth, and often, though not always, orange coloring of the ventral part of its body (Fig. 5.32). *Piranha caju* literally means cashew-fruit *piranha*, which is a reference to the similarity in orange coloring between ripe cashew fruit and the fish. Unless another species is involved, however, *Serrasalmus nattereri* individuals are also deep purple. It is one of the most abundant predators on the Rio Madeira floodplain, and I would estimate that it is at least as common as the *tucunaré* (*Cichla ocellaris*, Cichlidae), the most important piscivorous species in the commercial fisheries. The *piranha caju*, however, is highly destructive of fishing gear, tearing gillnets into shreds with its razor-like teeth; a school of them, when encircled in a seine, can annihilate this expensive piece of gear in a few seconds. Because there is no way of catching large numbers of them economically, they are relatively little exploited in the commercial fisheries. Their nasty habit – from a fisherman's viewpoint – of attacking hooked or gillnetted fishes, makes them a constant but unavoidable nuisance in the commercial fisheries of the Rio Madeira floodplain.

Prochilodontidae

Curimatá (*Prochilodus nigricans* Agassiz, Prochilodontidae)

Prochilodus nigricans is a silvery fish, reaching about 50 cm in standard length, and it one of the most abundant species in the Rio Madeira basin (Fig. 5.33). It lacks jaw teeth, but has protrusible lips that is uses to remove detritus from the bottom

Fig. 5.30 The *pacu mafurá* (*Myleus* sp., Characidae). About 22 cm standard length.

Fig. 5.31 A. The *sardinha comprida* (*Triportheus elongatus*, Characidae). About 17 cm standard length. B. The *sardinha chata* (*Triportheus angulatus*). About 12 cm standard length.

Fig. 5.32 The *piranha caju (Serrasalmus nattereri*, Characidae). About 21 cm standard length.

Fig. 5.33 The *curimatá (Prochilodus nigricans*, Prochilodontidae). About 27 cm standard length.

and other substrates. It is found in a variety of habitats during the year, including flooded forests, floodplain lakes, rainforest streams, and river channels. Along with other microphagous feeding fishes, namely, *Semaprochilodus* and several species of curimatids, *Prochilodus nigricans* begins its upstream migration in the Rio Madeira two or three months before most of the other migratory characins. *Semaprochilodus* species are usually the first fishes to descend the clearwater tributaries, but they are followed shortly thereafter by *Prochilodus nigricans* (usually in late April in the middle and upper Rio Madeira). The first annual migrations of *Prochilodus nigricans* appear to be relatively modest in comparison to the *Semaprochilodus* species, and this may in part account for the fact they, unlike the latter, display large-scale migrations during the low water period when, as fisheries data show, they are most captured (Fig. 5.34). The low catches of *Prochilodus nigricans* in December and January – the time when these fishes descend the tributaries to spawn in the Rio Madeira – is largely due to the fact that fishermen direct their effort to more valuable species, especially the *jatuarana* (*Brycon* sp., Characidae). The *curimatá* is considered a second class food fish, and in the 1970'a much of the Rio Madeira catch of the species was exported to Rio Branco, Acre.

Jaraqui Escama Grossa (*Semaprochilodus theraponura* Schomburgk) and *Jaraqui Escama Fina* (*Semaprochilodus taeniurus* Valenciennes, Prochilodontidae)

Fishes of the genus *Semaprochilodus* are characterized by their elongate bodies, toothless jaws, protrusible lips, and striped caudal, anal, and dorsal fins (Fig. 5.36). Their fins display black bars on an attractive yellow background. The body itself is silvery. The *jaraqui* are strong swimmers and, when frightened, can jump several meters out of the water. They are detritus feeding fishes, and use their protrusible lips – which are endowed with weak labial teeth – to remove fine particles from submerged substrates such as tree trunks, limbs, and sunken wood. In the Rio Madeira region, *Semaprochilodus theraponura* reaches about 40 cm in standard length, while *Semaprochilodus taeniurus* is somewhat smaller. The two species are easily distinguished by the number of lateral line scales (LLS); *S. theraponura* has an average of about 49-51 LLS and *S. taeniurus* has around 69 LLS.

Jaraqui are found mostly in the tributary systems of the Rio Madeira, whereas *Prochilodus nigricans*, a quite similar detritus feeding fish, dominates the floodplain of the principal river. The genus *Semaprochilodus* has not been reported above the first of the Rio Madeira rapids, and Bolivian and Brazilian fishermen exploiting the Rio Beni, Rio Mamoré, and Rio Guaporé say it is not in those rivers. Several residents of the Teotônio cataract of the Rio Madeira independently told me that at least two times in the last twenty years schools of *jaraqui* appeared at the rapids, and soon after there were large die-offs of these fishes that became strewn on the beaches and caused a fetid mess. Though this information is circumstantial, along with the known distribution, it does suggest that there are physical and/or biological factors that prevent fishes of the genus *Semaprochilodus* from passing the Rio Madeira rapids.

Jaraqui are the first fishes to migrate out of the flooded forests of the tributaries subsequent to entering these feeding habitats after spawning at the beginning of the annual inundation, and then move down the affluents in large schools and then up the Rio Madeira until entering another river farther upstream. In the three years for which accurate catch data are available, these migrations took place at or just after the height of the flood, that is, between about mid-March and early April (Fig. 5.35). Fishermen report that *jaraqui* schools migrate mostly during the waxing moon, and catch data indicate that this is true. It should also be pointed out, however, that most commercial fishermen desist from procuring *jaraqui* during the waning moon phases because their experience has taught them of the unproductiveness of fishing during this part of the month; therefore catch data are biased towards the waxing phases. Fishermen attempt to catch the *jaraqui* in the rivermouths of the tributaries that the fishes are descending, but if a full cargo is not made here, they follow them upstream until the schools enter another affluent.

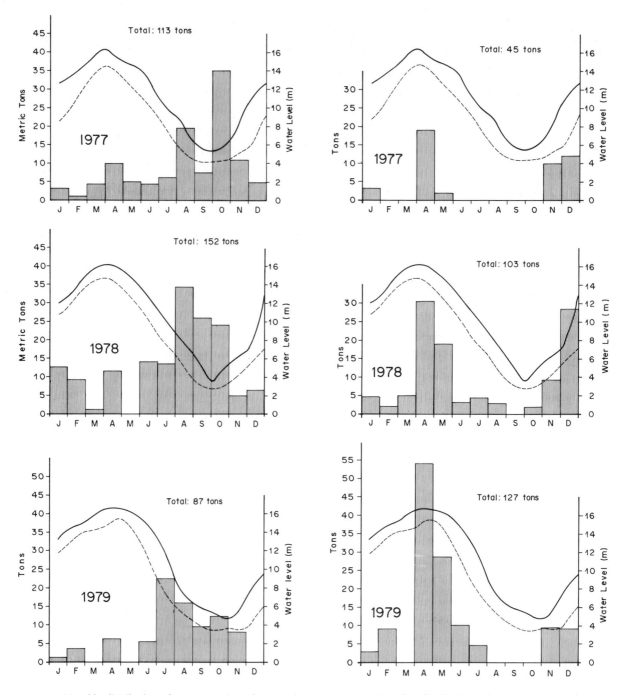

Fig. 5.34 Monthly distribution of annual catches of *curimatá* (*Prochilodus nigricans*, Prochilodontidae) by the Porto Velho fleet in the Rio Madeira in 1977, 1978, and 1979. The seasonal distribution of catches reflect clearly the relation between water level and the migratory behavior of the *Prochilodus nigricans*. In 1979, the exploitation of spawning *curimatá* was prohibited and thus there were no catches of the species in December. The solid and broken lines represent, respectively, the monthly high and low water levels.

Fig. 5.35 Monthly distribution of annual catches of *jaraqui* (*Semaprochilodus theraponura* and *taeniurus*, Prochilodontidae) by the Porto Velho fishing fleet in the Rio Madeira in 1977, 1978, and 1979. Note that the largest catches in all three years were made near the height of the floods. The solid and broken lines represent, respectively, the monthly high and low water levels.

Fig. 5.36 A. The *jaraqui escama grossa (Semaprochilodus theraponura*, Prochilodontidae); about 22 cm standard length. B. The *jaraqui escama fina (Semaprochilodus taeniurus)*; about 26 cm standard length.

In the upper Rio Madeira region, *Semaprochilodus theraponura* appears to be more abundant than *S. taeniurus*, and the latter is often captured in small quantities mixed with the former. A third cryptic species, the *jaraqui açú*, is also reported by fishermen, but I have not seen it in the Rio Madeira region, though its existence in the Rio Negro has been verified (it may be a hybrid between *theraponura* and *taeniurus*). During the low water season, *jaraqui* are only very rarely encountered in the Rio Madeira. When river level begins rising relatively rapidly in November and December, schools of *jaraqui* descend the tributaries to spawn in the turbid Rio Madeira; they appear to be the first of the migratory characin species to leave the affluents to spawn in the main river. Although fishermen could make larger catches than they do at this time, they are often hesitant to do so because of the relatively low market price of *jaraqui* in Porto Velho.

Curimatidae

Branquinha Chora (*Curimata latior* Spix), *Branquinha Cabeça Lisa* (*Curimata altamazonica* Cope), *Cascudinha* or *Chico Duro* (*Curimata amazonica* Eigenmann and Eigenmann) and *Branquinha Comum* (*Curimata vittata* Kner, Curimatidae)

Food fishes of the genus *Curimata* command little

respect for piscine beauty, and probably even less on the gastronomical score, but their abundance saves them from total disregard by commercial fishermen (Fig. 5.37). They are silvery fishes attaining, depending on the species, between about 15 and 25 cm in standard length when adult. All of the species are toothless when adult, and feed on detritus. Very little is known in detail about the kind of detritus they eat, or the microhabitat preferences of the individual species.

Because little fishing effort goes into catching curimatids, I am unable to say much about the migration patterns of the individual species. All of the species mentioned above, however, are found in the clearwater tributaries of the Rio Madeira and also on the floodplain of the main river. All four species in the commercial fisheries also descend the clearlwater tributaries to spawn in the Rio Madeira at the beginning of the floods. Schools of *Curimata latior* and *altamazonica* also appear in the Rio Madeira soon after the *Semaprochilodus* migrations at the height of the annual flood, but are not exploited by commercial fishermen. During the low water period, schools of all four species are among the most commonly encountered fishes in the Rio Madeira. The curimatids are usually only captured when other food fishes cannot be found and when ice is getting low.

Anostomidae

Aracu Botafogo (*Schizodon fasciatus* Spix, Anostomidae)

Schizodon fasciatus is a torpedo-shaped fish, reaching about 40 cm standard length, and is most characterized by its four vertical black hands and black spot between the cuadal fin and peduncle (Fig. 5.38). It has small, multi-cusped teeth, and is known to feed on leaves, fruit, and algae (in a clearwater tributary of the Rio Madeira studied by me), though Mendes dos Santos (1979) has shown convincingly that it feeds heavily on algae and macrophyte roots in the Rio Solimões floodplain. In the Rio Madeira, *Schizodon fasciatus* is captured mostly during the low water season when it is migrating upstream, and to a lesser extent during the spawning period when it is transiting the tributary mouths.

Pião (*Rhytiodus argenteofuscus* Kner and *Rhytiodus microlepis* Kner, Anostomidae)

Fishes of the genus *Rhytiodus* are the most elongated anostomids, and their common name *pião*, or toy-top, appears to be an allusion to their conical heads (Fig. 5.38). They reach at least 35 cm in standard length and are found in a wide variety of habitats, and often the two species are caught together in the same schools. They have fine, slightly cusped teeth and are known to feed on algae and macrophytes (Mendes dos Santos 1979). In color they are not very exciting, but usually have some mixture of dirty yellow with dark mottlings. They are considered second class food fishes in the Rio Madeira region, and are usually only exploited when other species cannot be found.

Hemiodontidae

Orana (*Hemiodus* spp., Hemiodontidae)

There are two or three species of *Hemiodus* captured in the Rio Madeira fisheries, but the taxonomy of the group is confusing and no reliable scientific names can be given at this time (Fig. 5.39). They are streamlined fishes with small mouths and only have teeth in the upper jaws. Their teeth, furthermore, are feeble and more or less cuspidate, but little is known of their feeding behavior. If the several species are taken together, they appear to have a relatively high biomass in the Rio Madeira, but they are considered second class food fishes and command only low prices in the Porto Velho Market.

Erythrinidae

Traíra (*Hoplias malabaricus* Bloch, Erythrinidae)

Hoplias malibaricus, or *traíra* as Brazilians call it, is

Fig. 5.37 A. The *branquinha cabeça lisa* (*Curimata altamazonica*, Curimatidae); about 18 cm standard length (SL). B. The *branquinha chora* (*Curimata latior*); about 20 cm SL. C. The *branquinha comum* (*Curimata vittata*); about 19 cm SL. D. The *cascudinha* or *chico duro* (*Curimata amazonica*); about 13 cm SL.

Fig. 5.38 A, B, E. Different colormorphs of the *aracu botafogo* (*Schizodon fasciatus*, Anostomidae). C, D. Unindentified anostomids. F. The *pião* (*Rhytiodus microlepis*, Anostomidae). All of the fishes shown above were removed from the same school which was captured in a beach area of the Rio Madeira during the low water period. Mimicry is suggested.

a vicious looking predator noted for its cylindrical body, rounded caudal fin, and relatively large, obliquely angled mouth that supports large canine-like jaw teeth (Fig. 5.40). The *traíra* is a common fish along the shorelines of floodplain lakes and lagoons, but because it lives solitarily or in small groups it is difficult to catch in large numbers. Most of the *traíra* catch in the Rio Madeira commercial fisheries is taken during the low and rising water period when the species becomes concentrated along the shores. The fish is killed mostly with gigs at night.

Cichlidae

Tucunaré (Cichla ocellaris Bloch and Schneider, Cichlidae)

Cichla ocellaris is the largest cichlid in the Amazon Basin, and also one of the favorite food and game fishes (Fig. 5.41). It grows to at least 75 cm in standard length and 12 kg in weight in the Rio Negro, but I have not seen any of this size in the Rio Madeira region. It is usually olivaceous to brown in color, with yellow to orange pigment on the lower fins and below the opercular opening. In mid-body it can have three black blotches or vertical bands, and there is always an ocellus on the upper and anterior part of the caudal fin. During the breeding season, the male has a large lump on its upper side anterior to the dorsal fin. The species is piscivorous, and found mostly on the floodplains. Most of the *tucunaré* catch of the Rio Madeira fisheries comes from the Cuniã floodplain, about 40 km from Porto Velho, but since about 1979, imports from the Rio Guaporé have exceeded the local yields. The techniques used to capture *tucunaré* were discussed in detail in Chapter 3.

Cará-açu (Astronotus ocellatus Spix, Cichlidae)

Astronotus ocellatus is the 'oscar' of aquarium buffs of the English speaking world (Fig. 5.42). Its Amazon name, *cará-açu*, means big *cará*; *cará* is the vernacular generic name for all the deep-bodied cichlids in the Amazon. *Astronotus ocellatus* reaches about 30 cm in standard length and in nature its body is olivaceous to brown in color with dark, and often thick, vertical stripes. Its outstanding feature, however, is its beautiful caudal ocellus which is framed in a bright red circle. Its feeding behavior has not been studied in detail, though it is known to eat insects and small fishes. It is a floodplain fish, and in the Rio Madeira region it is captured mostly with gigs at night. Like *Cichla ocellaris*, it is considered a first class food fish.

Osteoglossidae

Pirarucu (Arapaima gigas Cuvier, Osteoglossidae)

The *pirarucu* is one of the largest freshwater fishes in the world and reaches at least three meters in length and 150 kg in weight (Fig. 5.43). The fish has been exploited intensively for at least a century in the Amazon basin (Veríssimo 1895) and traditionally the flesh was salted and much of it exported to Northeastern Brazil and elsewhere. The vernacular name is taken from *lingua geral* and means red fish (*pira* = fish and *urucu* = red), an allusion to the often bright red scales that cover its body. In the Rio Madeira basin, the *pirarucu* is found in both the floodplain of the principal river and the floodplains of the clearwater tributaries. It is not found above the Rio Madeira rapids and appears to go no farther than the cataracts of the large rightbank tributaries.

In the Rio Madeira basin the *pirarucu* has been over-exploited, though there is still intensive fishing effort directed towards it. In the local folklore, it iione of the most prestigious fishes that a man can kill, and there is almost a tacit agreement that the harpooning of a *pirarucu* confers the right of manhood on youth that are recruited into subsistent and commercial fisheries. Traditionally the *pirarucu* was taken mostly with the harpoon, as the fish is an air-breather and must surface every few minutes, and thus is vulnerable at or near the surface. As large fish have disappeared because of over-exploitation, gillnets have become more important in catching *bocó*, or immature individuals under about 1.25 m and 40 kg. Largely because of

Fig. 5.39 The *orana* (*Hemiodus* sp., Hemiodontidae). About 25 cm standard length.

Fig. 5.40 The *traíra* (*Hoplias malabaricus*, Erythrinidae). About 30 cm standard length.

Fig. 5.41 The *tucunaré (Cichla ocellaris*, Cichlidae). About 30 cm standard length.

Fig. 5.42 The *cará-açu (Astronotus ocellatus*, Cichlidae). About 22 cm standard length.

Fig. 5.43 Drawing of a *pirarucu* (*Arapaima gigas*, Osteoglossidae) from Franz Keller's *The Amazon and Madeira Rivers* (1874).

gillnets, the *pirarucu* is now seriously threatened with extinction in the Rio Madeira basin, especially in the clearwater tributaries where populations were probably never very large in the first place.

Aruanã (*Osteoglossum bicirrhosum* Vandelli, Osteoglossidae)

Osteoglossum bicirrhosum is an elongated, greatly compressed fish, with a huge mouth and large scales (Fig. 5.44). It reaches at least one meter in length. The *aruanã* is mostly a floodplain fish, and though omnivorous, it has a preference for insects and spiders which it takes at the surface and usually near to the shore. It is captured mostly during the low water season with gillnets (to which it is highly vulnerable), castnets in the shallow lakes and lagoons, bows-and-arrows, and gigs at night. Though not a favored food fish, the *aruanã* is somewhat important in the folklore of the region as it is one of the few species that *caboclo*, or riparian peasant, women are allowed to eat during postpartum convalescence.

Sciaenidae

Pescada (*Plagioscion squamosissimus* Haekel, Sciaenidae)

There are several species of croakers, or drums, in the Rio Madeira basin, but only *Plagioscion squamosissimus* is of commercial importance (Fig. 5.45). It is a silvery fish with a large, oblique mouth, and best known for the sounds (produced by drumming muscles on the swimbladder) it produces. When adult it is mostly piscivorous, but also feeds on shrimp and crab. Small schools of *pescada* are encountered in the Rio Madeira during the low water season, but they do not show any distinct migration patterns as do the characins and catfishes. Seine fishermen occasionally catch them below the Teotônio and Santo Antonio cataracts, and the croakers may move upstream to these areas during low water to feed on the smaller fishes that become concentrated here. The other commercial fishery for the croaker in the Rio Madeira is in the mouths of some of the clearwater tributaries, especially the Rio Jamari and Rio Machado. The croakers become abundant near the confluences of the affluents and the principal river during the low water period, and this, again, appears to be for feeding on prey entering and exiting the clearwater rivers. In the tributary mouths, croakers are captured with handlines. The fisherman 'sounds' for the *pescada* by allowing his line and baited hook to sink vertically to various depths; at each determined depth the baited hook is made to bob up-and-down in an attempt to attract a croaker that might be near. When a croaker finally takes the baited hook, the fisherman marks his line and continues to fish at that depth where groups or schools of *pescada* reside. The croakers appear to be sensitive to depth, even in relatively shallow waterbodies such as Amazon rivers and floodplain lakes. The *pescada* is considered one of the best food fishes of the region, but is not very important in the total catch.

Clupeidae

Apapá (*Pellona castelnaeana* Valenciennes, Clupeidae)

Pellona castelnaeana is a large predatory clupeid characterized by its yellow body and small, upturned mouth (Fig. 5.46). It feeds in the crespuscular hours at which time it can be seen attacking small insectivorous fishes at the surface. Fishermen use handlines with the hooks baited with pieces of fish; the line is retrieved in quick, short jerks, fast enough to keep the bait at the surface. When the predator strikes, it is hooked and pulled into the canoe. Small schools of *Pellona castelnaeana* occasionally appear at the Rio Madeira rapids and are there captured with seines. The *apapá* is considered a second or even third class food fish, and thus little effort goes into catching it.

Fig. 5.44 The *aruanã* (*Osteoglossum bicirrhosum*, Osteoglossidae).

Fig. 5.45 The *pescada* (*Plagioscion squamosissimus*, Sciaenidae). About 28 cm standard length.

Fig. 5.46 The *apapá* (*Pellona castelnaeana*, Clupeidae).

CHAPTER 6

Problems and prospects

The way in which the fisheries of the Rio Madeira valley will be used or abused will depend, in large part, on decisions made within the broader framework of the economic development of the Amazon basin. Of most importance will be the degree to which the floodplains are deforested for agricultural schemes; large-scale deforestation would probably destroy the fisheries as they are presently known. Other economic decisions that will affect the fisheries will be: the need, as perceived by the Brazilian and Bolivian governments, for large dams to supply badly needed energy to a region that is booming in population and paying high thermoelectric bills; the future availability of alternative animal protein sources for the growing urban centers, and; the ability of authorities to implement a workable management program under tight budgets in a region that is characterized by a high rate of illiteracy, poverty but rising expectations, and serious malnutrition. In view of the above points, the main purpose here is to summarize what is known about the natural history of the fisheries so that these conclusions are at least available for weighing against those future programs that might affect the piscine resource in a negative way.

Relative productivity of the Rio Madeira fisheries

Yields in comparison to floodplain area
Ideally, providing there is the need for all animal protein potentially available, fisheries should be managed from the standpoint of exploiting the communities as heavily as possible without destroying them. This concept is based on the generally accepted hypothesis that fish communities produce a constant surplus that can be removed by fisheries (Lowe-McConnell 1975; Welcomme 1979). Welcomme (1979), however, has pointed out that it is extremely difficult to estimate maximum sustained yields for river fisheries, as they are usually based on multispecies communities whose annual abundance is largely determined by flood levels. In this view, there is too much variation in annual abundances to arrive at an accurate estimate of constant surplus production. In absolute, quantitative terms, I would certainly agree that the maximum sustained yield concept is probably a difficult one to apply rigidly to the management of river fisheries, but I nevertheless feel that correlations made between effort and yield over a series of years (a minimum of five years is suggested for the Amazon) will indicate the general limits of surplus fish production that can be harvested without destroying the fisheries. Even if the limits are estimated too low, it would appear best to underexploit than overexploit the fisheries until more is known about them.

Welcomme (1979) reviewed the available data on the average fisheries yield of floodplain rivers, and arrived at the conclusion that it is somewhere between 40 and 60 kg/ha of floodplain/yr. The Rio Madeira and the lower courses of its tributaries (excluding the Beni-Mamoré-Guaporé region) appear to have about 150,000 hectares of floodplain.

The total estimated commercial and subsistence catch of the Rio Madeira in 1977 was at least 7,800 metric tons (Bayley 1978; Goulding 1979), and this gives a yield of 52 kg per hectare when divided into the total floodplain area; it should be remembered that total floodplain area is used only as an index for comparison, and does not necessarily indicate the role of the river channel itself in fish production. The Rio Madeira yields, when compared to total floodplain area, fall well within the limits cited by Welcomme (1979) for most tropical rivers. The comparative data cited above, and more importantly perhaps, the general decline since 1976 in total annual catches and catch per unit of effort as evidenced by the Porto Velho fleet, suggest that the Rio Madeira does not have a large untapped fish resource. In short, not much more should be expected from the Rio Madeira fisheries in terms of total potential catch.

Introduction of exotic species

Exotic fish species have been introduced around the world, and in many cases with the idea of increasing piscine productivity of the natural systems. In terms of historical zoogeography uninfluenced by man, the Amazon basin can probably claim the world's most natural fish fauna. The idea of the successful exotic, however, is currently making its rounds in the Amazon region. At the top of the list of tropical fish exotics are tilapia, and at least one species (*Sarotherodon*? sp., Cichlidae) has been introduced into Rondônia, and much of the Amazon basin for that matter. Most tilapia imported into the Amazon have been used for fish culture purposes, though there have been numerous escapes, especially during the flooding season when poorly constructed impoundsments placed across rainforest streams ruptured and the exotic cichlids fled to the rivers. Though tilapia have entered the rivers, there is no evidence to date that they have established natural populations, but it should also be pointed out that the African cichlids are very similar in appearance to some of their American counterparts, and it is highly unlikely that local residents would recognize them as foreign. The main reason that tilapia have been chosen by Amazonian fish culturalists is that there is a relatively large literature on them, whereas information on the native species is scarce and mostly esoteric. There is also the belief that exotics will fill-up uninhabited niches, but it should be reiterated here that the Amazon basin has the most diverse freshwater fish fauna in the world, and includes a large number of species of all the major trophic groups found in other tropical regions. The most learned opinions now feel that introducing exotic species for food fishes calls for careful research before doing so, as introductions may cause more problems than they attempt to solve (Lowe-McConnell, 1975).

Extermination of fish predators

It has been suggested in both the literature (Meschkat 1960) dealing with Amazonian fisheries and by local fish authorities whom I have interviewed, that the extermination of piscivorous predators, such as dolphins, caimans, and large water birds, might increase fish productivity. The facile idea, simply stated, is that these predators consume the same fishes that could also be utilized by man, and hence the piscivorous mammals, reptiles, and birds are competing with the fisheries. This seemingly logical belief, however, fails to consider the overall role of predators in ecosystems. Most work in this regard on tropical fishes has been done in African waterbodies, and it has been generally concluded that piscivorous predators are not harmful to the fisheries, and that they may even have a beneficial role in nutrient recycling (Bowmaker, 1963; Welcomme 1979). Fittkau (1970, 1973) hypothesized that caimans may contribute significantly to the nutrient balance in the mouthlake areas of nutrient-poor rivers in the Amazon, and that these salts would be important for building up a food chain, through phytoplankton and zooplankton, for fish fry nourished in these waterbodies. He also reported that there have been declines in catches in areas where caimans have been destroyed, though he failed to state exactly where. Needless to say, in an area such as the Amazon with so many nutrient-poor river systems, the role of all types of animals in nutrient recycling deserves more attention.

Direct management of fisheries

There are several possible ways to manage fisheries directly, of which closed seasons, gear restrictions, restrictions on numbers of fishermen or fishing boats, and reserved areas, are the most commonly attempted. Each of these will be discussed within the Amazonian biological and cultural framework.

Closed seasons

The distribution of monthly catches shows that the Rio Madeira fisheries have a built-in closed season, as is generally true for most floodplain river fisheries (Welcomme 1979). When the fishes are spread out in the flooded forests during the annual inundations, most species are very difficult to catch in large quantities. This closed season imposed by nature lasts for about five or six months for most of the commercial species in the Rio Madeira fisheries. As discussed earlier, the widespread use of large-meshed gillnets makes the deep-bodied fishes of the genus *Colossoma* vulnerable even during the floods when they are in the flooded forests. The smaller fishes are also vulnerable to gillnets, but smaller meshes are avoided because of *piranhas* that can render this type of fishing expensive. As a community, then, it can be concluded that most of the commercial species are offered protection by a naturally built-in closed season, or the period of annual floods.

In the Amazon, fish authorities have on occasion prohibited the exploitation of migratory characins during their spawning runs at the beginning of the floods. This management strategy is based on the belief that the exploitation of fishes during the spawning period has a greater negative effect on populations than comparable captures made at other times of the year. In all objectivity, there is not enough information available for Amazon fisheries ecology to know if this is true or not. My own feeling is that it makes little difference whether the fishes are taken during the spawning period or during their dispersal migrations (see Chapter 5), as the net result is the same: the removal of potential reproducers of the population.

Gear restrictions

It is highly unlikely that authorities will be able to control the types of fishing gear that are used in the Amazon region. To police fishing operations in an area so large would demand the allocation of financial resources beyond the present capacities of the already lean budgets of fish authorities, not to mention the lack of trained personnel to implement such a program. The diffusion of gillnets is taking place so rapidly in the Amazon, that virtually every large fishing boat and riparian settlement has them. The effectiveness of gillnets, especially for the deep-bodied fishes, makes it highly unlikely that commercial and subsistence fishermen would give them up, even if they were prohibited. There has been some effort to control mesh sizes, but *piranhas*, in the case of the Amazon, may be as effective at this as would be fish authorities. In the Rio Madeira region, small-meshed gillnets are of almost no importance in the commercial fisheries. To limit the length and mesh size of seines is probably of no great value in the Rio Madeira fisheries. Nearly all of the commercial catch taken with seines consists of mature fishes, and in four years of investigation of the Rio Madeira I found very little evidence of the serious or wanton destruction of smaller size classes due to small-meshed seines.

Reserved areas

From a biogeographical perspective, the most important point to be made about the Rio Madeira fisheries is that the majority of food fishes utilize two radically different hydrochemical environments during their lives. The young of the migratory characins, overall the most important commercial taxa, are nourished in Rio Madeira floodplain lakes and lagoons that are injected annually with turbid waters charged with nutrients washed out of the Bolivian Andes. Later in their lives, the exact times still unknown, many and probably most of the food fish characins migrate out of the Rio Madeira floodplain waterbodies and move into the clearwater and blackwater tributaries where, as adults, they find most of their food in the flooded forests during the annual inundations. The hypothetical migrations were discussed in Chapter 5. Some of the large catfishes appear to have even more extensive migrations, and some species may begin their upstream journey in the Rio Amazonas and go as far as the large savanna areas of eastern Bolivia. In a sense, the Rio Madeira is used as a conduit by the large siluroids to move from one large floodplain area to another, more than a thousand kilometers distant. The data presented clearly indicate that nearly all of the characins and catfishes of commercial importance have life cycles that embrace large geographical areas, and this point must always be kept in mind when considering management strategies for the fisheries. To protect small, isolated areas, may be to almost no avail in terms of preserving the fisheries of the region in general. In the case of protecting the migratory characins, the most important food fishes, reserved areas should include entire tributary systems, or at least the greater part of some of them, and Rio Madeira floodplain areas (especially in the lower course of the river) where alevins are nourished and from where they are recruited into the fisheries.

The large catfishes appear to present political problems as their life cycles embrace both Brazilian and Bolivian waters. Their protection may eventually call for international cooperation.

Environmental protection

Unlike anywhere else in the world, the flooded forests of the Amazon play a key role in fish nutrition (Goulding 1980). The fact that the food chain leading to perhaps 75 percent of the total commercial catch begins in flooded forests suggests that these floodplain habitats must be protected if the fisheries are to be preserved on their present scale. In the nutrient poor systems, the blackwaters and clearwaters, it is highly unlikely that there is an alternative food chain to support large numbers of fishes once the floodplain forests are destroyed. As was shown in this study, the nutrient poor tributaries of the Rio Madeira are where most of the commercial catch is nourished, and thus forest reserves should include both the principal river and its tributaries.

Since there are plans for deforesting large areas of the turbid river floodplains for rice culture and other agricultural activities, it should be reiterated here that these areas are the nursery grounds for most of the food fishes, and furthermore, for many of the fishes who later spend their lives in the clearwater and blackwater tributaries. Destruction of young fish in the turbid water floodplain nursery habitats would also mean the destruction of fishes that would also later inhabit the nutrient poor tributary systems of the main river. Probably the major threat to young fishes with the modification of floodplains for agriculture would be the heavy use of insecticides. Since floodplain waterbodies become very restricted during the low water period, the young fishes, and many older ones as well inhabiting the floodplains, become very concentrated. These restricted waterbodies would act as insecticide traps collecting the poisons from groundwater and run-off in the surrounding agricultural areas. In short, to protect the commercial fish fauna, not only must the flooded forests be preserved, but water quality must also be maintained.

Bibliography

Abreu, J.C. 1930. Caminhos Antigos e Povoamento do Brasil. Livraria Briguiet, Rio de Janeiro.

Anonymous, 1977. Potencial hidroeléctrica do Madeira chega a 13 milhões de KW. Tribuna (Porto Velho). 29 de setembro de 1977, p. 2.

Bayley, P.B. 1973. Studies on the migratory characin, *Prochilodus platensis* Holmberg, 1889 (Pisces: Characodei) in the R. Pilcomayo, South America. J. Fish. Biol. 5: 25-40.

Bayley, P.B. 1978. Fishery yield from the middle and upper Amazon in Brazil: a comparison with exploitation in African rivers. Manuscript presented to International Workshop on Comparative Studies in Freshwater Fisheries, Pallanza, Italy, 4-8 September 1978.

Bayley, P.B. 1979. The limits of limnological theory and approaches as applied to river-floodplain systems and their fish production. In Welcomme, R.L. (Ed.). Fishery Management in Large Rivers. FAO Technical Paper No. 194, Rome: 23-26.

Beurlen, K. 1970. Geologie von Brasilien. Beiträge zur regionale Geologie der Erde. Berlin-Stuttgart.

Bowmaker, A.P. 1963. Comorant predation on two central African lakes. Ostrich 34(1) : 2-26.

Bowman, J. 1913. Geographical aspects of the new Madeira-Mamoré railroad. Bull. Amer. Geogr. Soc. 45: 275-281.

Britski, H. In Press. *Merodontotus tigrinus*, um gênero e espécie novo do Sorubiminae na bacia amazônica. Papeis Avulsos Mus. Zool. S. Paulo.

Carvalho, F.M. 1979. Estudo da alimentação, desenvolvimento dos ovários e composição química de *Hypophthalmus edentatus e Potamorhina pristigaster*. Master's Thesis, Instituto Nacional de Pesquisas da Amazônia, Manaus, Amazonas.

Coget. 1978. População de Porto Velho (Unpublished data supplied by Coordenação de Geografia e Estatística, Porto Velho, Rondônia).

Cortesão, J. 1958. Raposo Tavares e a Formação Territorial do Brasil. Ministério da Educação e Cultura, Rio de Janeiro.

Craig, N.B. 1907. Recollections of an I11-Fated Expedition to the Headwaters of the Madeira River in Brazil. J.B. Lippincott Co., Philadelphia.

Denevan, W.M. 1966. The aboriginal cultural geography of the Llanos de Mojos of Bolivia. Ibero-Americana. No. 48.

Ferreira, A.R. 1972. Viagem Filosófica pelas Capitanias do Grão-Pará, Rio Negro, Mato Grosso e Cuiabá. Conselho Federal de Cultura, Rio de Janeiro.

Fittkau, E.J. 1970. Role of caimans in the nutrient regime of mouth-lakes of Amazon affluents (a hypothesis). Biotropica 2(2): 138-142.

Fittkau, E.J. 1973. Crocodiles and the nutrient metabolism of Amazonian water. Amazoniana 4 (1): 101-133.

Fittkau, E.J. 1974. Zur ökologischen Gliederung Amazoniens. I. Die erdgeschichtliche Entwicklung Amazoniens. Amazoniana 5(1): 77-134.

Gibbon, L. and Herndon, W. 1854. Exploration of the valley of the Amazon. United States Government, Washington, D.C. Vol. 2.

Gibbs, R.J. 1967. The geochemistry of the Amazon river system: Part I. The factors that control the salinity and the composition of the suspended solids. Geol. Soc. Amer. Bull. 78: 1203-1232.

Giugliano, R., Shrimpton, R., Arkcoll, D., Guiugliano, L.G. and Petrere, M. 1978. Diagnóstico da realidade alimentar e nutricional do Estado do Amazonas, 1978. Acta Amazonica 8(2): Suplemento 2.

Goulding, M. 1979. Ecologia de Pesca do Rio Madeira. Conselho Nacional de Desenvolvimento Científico e Tecnológico e Instituto Nacional de Persquisas de Amazônia, Manaus.

Goulding, M. 1980. The Fishes and the Forest: Explorations in Amazonian Natural History. University of California Press, Los Angeles.

Gourou (1950). Observações geográficas na Amazônia. Rev. Bras. Geogr. 9(3): 355-408.

Grabert, H. 1967. Sobre o desaguamento natural do sistema fluvial do Rio Madeira desde a construção dos Andes. Atas do Simpósio sobre a Biota Amazônica 1: 209-214.

Greenwood, P.H., Rosen, D.E., Weitzman, S.H. and Myers, G.S. 1966. Phyletic studies of Teleostean fishes, with a provisional classification of living forms. Bull. Amer. Mus. Nat. Hist. 131(4): 341-455.

Hidrologia. 1974-1979. Cotas fluviométricas e descargas médias. (Data supplied by Centro de Pesquisas de Recursos Minerais, Manaus).

Hugo, V. 1959. Desbravadores. Edição da 'Missão Salesiana de Humaitá', Humaitá (Amazonas). Vol 1.

IBGE. 1975a. Atlas de Rondônia. Instituto Brasileiro de Geografia e Estatística, Rio de Janeiro.

IBGE. 1975b. Estimativa da população residente nas regiões fisiográficas, unidades de federação, microregiões homogenias, áreas metropolitaneas, e municípios em 1º de julho de 1975. Instituto Brasileiro de Geografia e Estatística, Rio de Janeiro.

IBGE. 1977. Geografia do Brasil: Região Centro-Oeste, Vol. 4. Instituto Brasileiro de Geografia e Estatística, Rio de Janeiro.

IBGE. 1979. A Organização do Espaço na Faixa da Transamazônica. Convênio Fundação Instituto Brasileiro de Geografia e Estatística e Secretária de Planejamento de Presidência da República, Rio de Janeiro.

Ihering, R. von. 1929. Da Vida dos Peixes. Companhia Melhoramentos, São Paulo.

Junk, W.J. 1970. Investigations on the ecology and production-biology of the 'floating meadows' (Paspalo-Echinochloetum) on the middle Amazon. I. The floating vegetation and its ecology. Amazoniana 2: 449-495.

Junk, W.J. 1973. Investigations on the ecology and production-biology of the 'floating meadows' (Paspalo-Echinochloetum) on the middle Amazon. II. The aquatic fauna in the root-zone of floating vegetation. Amazoniana 4(1): 9-102.

Keller, F. 1874. The Amazon and Madeira Rivers. Chapman and Hall, London.

Klinge, H. 1967. Podzol soils: a source of blackwater rivers in Amazônia. Atas do Simpósio sobre a Biota Amazônica 3: 117-125.

Lowe-McConnell, R.H. 1975. Fish Communities in Tropical Freshwaters. Longman, London.

Menezes, N.A. 1970. Distribuição e origem da fauna de peixes de água doce das grandes bacias fluviais do Brasil. In Comissão Interestadual da Bacia Paraná-Uruguai (Ed.). Poluição e Piscicultura. Faculdade de Saúde Pública de USP, São Paulo: 73-78.

Mendes dos Santos, G. 1979. Estudo da alimentação, reprodução e aspectos da sistemática de Schizodon fasiatus Agassiz, 1829, Rhytiodus microlepis Kner 1859 e Rhytiodus argenteofuscus Kner 1859 do Lago do Janauacá – Am., Brasil. Master's Thesis, Instituto Nacional de Pesquisas da Amazônia, Manaus, Amazonas.

Meschkat, A. 1960. Report to the Government of Brazil of the fisheries of the Amazon region. FAO Report Nº 1305, Rome.

Myers, G.S. 1949. The Amazon and its fishes (Part V). A monograph on the piranha. The Aquarium Journal Feb.: 52-61.

Paixão, I.M.P. 1980. Estudo da alimentação e reprodução de Mylossoma duriventris Cuvier, 1818 (Pisces, Characoidei), Do Lago Janauacá, Am., Brasil. Master's Thesis, Instituto Nacional de Pesquisa da Amazônia.

Pearson, N.E. 1937. The fishes of the Beni-Mamoré and Paraguay basins, and a discussion of the origin of the Paraguayan fauna. Proc. Calif. Acad. Sci. 4(23): 99-114.

Petrere, M. 1978a. Pesca e esforço de pesca no Estado do Amazonas. I. Esforço e captura por unidade de esforço. Acta Amazonica 8(3): 439-454.

Petrere, M. 1978b. Pesca e esforço de pesca no Estado do Amazonas. II. Locais, aparelhos de captura e estatísticas de desembarque. Acta Amazonica 8(3): Suplemento 2.

Pires, J.M. 1974. Tipos de vegetação da Amazônia. Bras. Flor. 5(17): 48-58.

Portobras (1976-1979). Cotas fluviométricas (data supplied by Manaus, Amazonas office).

Prance, G.T. 1978. The origin and evolution of the Amazonian flora. Interciencia 3(4): 207-222.

Ribeiro, D. 1976. Uirá Sai à Procura de Deus. Editora Paz e Terra S/A.

Ringuelet, R.A., Aramburu, R.H. and Aramburu, A.A. 1967. Los Peces Argentinos de água Dulce. Comision de Investigacion Cientifica, La Plata.

Roberts, T. 1972. Ecology of fishes in the Amazon and Congo basins. Bull. Mus. Comp. Zool. Harvard 143(2): 117-147.

Rocque, C. 1968. Grande Enciclopédia da Amazônia. Amazônia Editora Ltda., Belém.

Roosevelt, T. 1914. Through the Brazilian Wilderness. Charles Scribner's Sons, New York.

Schmidt, G.W. 1973a. Primary production of phytoplankton in three types of Amazonian waters. II. The limnology of a tropical floodplain lake in Central Amazônia (Lago de Castanho). Amazoniana 4(2): 139-203.

Schmidt, G.W. 1973b. Primary production of phytoplankton in three types of Amazonian waters. III. Primary production of phytoplankton in a tropical flood-plain lake of Central Amazônia, Lago do Castonho, Amazonas, Brazil. Amazoniana 4(4): 379-404.

Sioli, H. 1967. Studies in Amazonian waters. Atas do Simpósio sobre a Biota Amazônica 3: 9-50.

Sioli, H. 1968. Hydrochemistry and geology in the Brazilian Amazon region. Amazoniana 1(3): 267-277.

Smith, N.J.H. 1979. A pesca no rio Amazonas. Conselho Nacional de Desenvolvimento Científico e Tecnológico e Instituto Nacional de Pesquisa da Amazônia, Manaus.

Thery, H. 1976. Rondônia: Mutations d'un Territoire féderal en Amazonie Brésilienne. Ph.D. Dissertation, Université de Paris.

Tomlinson, H.M. 1928. The Sea and the Jungle. The Modern Library New York.

Welcomme, R.L. 1979. Fisheries Ecology of Floodplain Rivers. Longman, London.

Wesche, R. 1978. A moderna ocupação agrícola em Rondônia. Rev. Bras. Geogr. 40(3/4): 233-247.

Veríssimo, J. 1895. A Pesca na Amazônia. Livraria Classica de Alves & Co., Rio de Janeiro.

Author index

Abreu, J.C., 17
Anonymous, 16

Bayley, P.B., 57, 59, 60, 65, 71, 118
Beurlen, K., 4
Bowmaker, A.P., 118
Bowman, 22
Britski, H., 84

Carvalho, F.M., 91
COGET, 24
Cortesão, J., 17
Craig, N.B., 22
Devevan, W.M., 4

Ferreira, A.R., 17, 22, 26
Fittkau, E.J., 4, 118

Gibbon, L., 23
Gibbs, R.J., 3, 5, 10
Giugliano, R. et al., 24
Goulding, M., 12, 24, 42, 43, 62, 72, 118, 120
Gourou, 4
Grabert, H., 4
Greenwood, P.H. et al., 73

Herndon, W., 23
Hidrologia, 1, 10
Hugo, V., 17

IBGE, 23, 24
Ihering, R. von, 78, 80

Junk, W.J., 12, 71

Keller, F., 18, 19, 23, 26, 27, 112
Klinge, H., 10

Lowe-McConnell, R.H., 117, 118

Menezes, N.A., 1
Mendes dos Santos, G., 106
Meschkat, A., 118
Myers, G.S., 19

Paixão, I.M.P., 98
Pearson, N.E., 1
Petrere, M., 24, 44, 52, 60, 65, 97
Pires, J.M., 4
Portobras, 10
Prance, G.T., 12

Ribeiro, D., 19
Ringuelet, R.A. et al., 78
Roberts, T., 73
Rocque, C., 43
Roosevelt, T., 19, 74

Schmidt, G.W., 71
Sioli, H., 5
Smith, N.J.H., 4, 10, 24, 44, 65

Thery, H., 23
Tomlinson, H.M., 22

Welcomme, R.L., 43, 63, 117, 118, 119
Wesche, R., 23

Veríssimo, J., 43, 54, 109

Index to scientific names

Amanoa, 100
Anodus, 69
Anodus elongatus, 39
Anostomidae, 29, 73, 106, 108
Arapaima gigas, 44, 45, 49, 50, 73, 74, 109–113
Astrocaryum jauary, 51, 52, 87, 89, 90, 97
Astronotus ocellatus, 44, 45, 46, 109, 110, 111

Bombacaceae, 12, 14
Brachyplatystoma, 80
Brachyplatystoma filamentosum, 28, 32, 38, 39, 40, 56, 74–76
Brachyplatystoma flavicans, 27, 28, 29, 30, 38, 56, 60, 72, 73–75, 84, 92
Brachyplatystoma juruense, 49, 84, 85
Brachyplatystoma sp., 45, 49
Brachyplatystoma vaillantii, 28, 76, 77
Brycon, 38, 39, 45, 52, 56, 60, 69, 72, 91–93, 94, 95, 96, 103

Callophysus macropterus, 28, 32, 84, 87, 88
Characidae, 38, 39, 42, 46, 49, 51, 56, 60, 73, 91–100, 101, 103, 118
Ceiba pentandra, 12, 14, 16
Cetopsidae, 29, 31
Cetopsis, 29, 31
Chironomus, 87, 91
Cichla ocellaris, 44, 45, 46, 48, 50, 56, 60, 72, 73, 100, 109, 110
Cichlidae, 44, 45, 56, 60, 72, 109, 110
Clupeidae, 113
Colossoma, 69, 70, 119
Colossoma bidens, 39, 45, 49, 52, 60, 98, 99
Colossoma macropomum, 38, 39, 42, 43, 45, 49, 51, 52, 56, 60, 96, 97–98
Curimata, 39, 54, 69, 72, 105–106, 107
Curimata altamazonica, 38, 39, 45, 60, 73, 105–106, 107
Curimata amazonica, 38, 39, 60, 105–106, 107
Curimata latior, 38, 39, 45, 60, 105, 107
Curimata vittata, 38, 39, 45, 60, 105, 107
Curimatidae, 39, 45, 105–106, 107

Doradidae, 73, 97–91, 92

Echinochloa, 12
Electrophorous electricus, 73
Ephermeroptera, 87
Erythrinidae, 43, 73, 105, 109, 110
Euphorbiaceae, 97, 98

Goslinia platynema, 23, 28, 29, 72, 73, 76–78
Hemicetopsis, 29
Hemiodus, 29, 69, 106, 110
Hevea brasiliensis, 51, 52, 97
Hevea spruceana, 51, 52, 97
Hoplias malabaricus, 44, 45, 106–109, 110
Hypophthalmidae, 73, 94
Hypophthalmus edentatus, 91, 94
Hypophthalmus perporosus, 91

Inia geoffrensis, Platanistidae, 42
Leporinus, 69
Leporinus friderici, 49
Licania longipetala, Chrysobalanaceae, 89, 91
Lithodoras dorsalis, 28, 89, 91, 92
Loricariidae, 45, 50, 91
Luffa, Cucurbitaceae, 51

Megaladoras irwini, 28, 87–89, 90, 91
Merodontotus tigrinus, 28, 84, 86
Mora paraensis, Leguminosae, 49
Myleus, 56, 98, 100, 101
Mylossoma, 56, 69, 70, 72, 98, 100
Mylossoma albiscopus, 39, 45, 60, 98, 99
Mylossoma aureus, 39, 45, 60, 98, 99
Mylossoma duriventris, 39, 45, 60, 98

Osteoglossidae, 44, 45, 48, 49, 73, 109–113, 114
Osteoglossum bicirrhosum, 44, 45, 48, 73, 113, 114
Oxydoras niger, 28, 87, 89

Palmae, 39, 87, 89, 97
Pareiodon, 29
Paspalum, 12
Paspalum repens, Gramineae, 91
Paulicea lutkeni, 28, 32, 33, 35, 38, 78–81
Pellona castelnaeana, 113, 115
Phractocephalus hemiliopterus, 28, 38, 83–84, 85
Pimelodella, 78
Pimelodus, 28, 78
Pimelodus blochii, 78, 79
Pinirampus pirinampu, 28, 84–86, 88
Plagioscion squamosissimus, 113, 115
Platynematichthys notatus, 83
Plecostomus spp., 45, 50, 91, 93
Prochilodontidae, 39, 45, 46, 47, 54, 56, 57, 60, 80, 100–104
Prochilodus, 38, 69, 70, 72
Prochilodus nigricans, 39, 45, 46, 47, 54, 56, 57, 60, 80, 100-104
Prochilodus platensis, 48
Prochilodus scrofa, 81
Podocnemis expansa, Pelomedusidae, 58
Pseudoplatystoma, 28, 80
Pseudoplatystoma fasciatum, 28, 56, 82–83
Pseudoplatystoma tigrinum, 28, 56, 81–82, 83
Pseudostegophilus, 29, 31

Pterodoras granulosus, 28, 87, 90
Pterygoplichthys, 50, 91

Rhytiodus, 69
Rhytiodus argenteofuscus, 39, 106
Rhytiodus microlepis, 39, 106, 108

Schizodon, 64
Schizodon fasciatus, 39, 106, 108
Sciaenidae, 113
Semaprochilodus, 38, 43, 56, 63, 69, 70, 72, 103, 106
Semaprochilodus taeniurus 39, 60, 69, 103–105
Semaprochilodus theraponura, 39, 45, 60, 69, 103-105
Serrasalmus nattereri, 45, 49, 100, 102
Serrasalmus rhombeus, Characidae, 56
Sarotherodon, 118
Sorubim lima, 28, 78, 79
Sotalia fluviatilis, Delphinidae, 42
Surubimichthys planiceps, 28, 84, 86

Trichomycteridae, 29, 31
Triportheus, 38, 49, 60, 69, 70, 72, 100, 102
Triportheus angulatus, 38, 39, 60, 100, 101
Triportheus elongatus, 38, 39, 60, 100, 101

Subject index

Acids, humic, 10
Acre, Territory of, 24
Agents, seed dispersal, 89, 91
Agriculture,
　modification of floodplains for, 120
　crop, 16
Air-breathers, 91, 109
Alevins, 12, 63, 70, 71, 120
　of the migratory characins, 70
　transfering energy to, 71
Agassiz, 19
Algae, 106
Amazon, 17, 26, 38, 54, 73, 73, 76, 78, 117, 118, 119
　Central, 38
　flooded forests of, 120
　southern highlands of, 19
　western, 25, 65
Amazonas, 25, 26, 60, 97
Americans, 22
Amerind, 4, 19, 26
Analyses, stomach content, 72, 73
Andean Cordillera, 5
Andes, 1, 4, 5, 8, 43, 119
　Bolivian, 1, 119
Anostomids, 60, 106, 108
Apapã, 113, 115
Aracu botafogo, 39, 106, 108
Aracu cabeça gorda, 39
Archers, 46, 47
　piscine, 46
Area(s),
　deforested, 120
　reserved, 119, 120
　savana, 4, 120
　woody shore (*pauzadas*), 97
Ariquemes, 24
Arpão, 49
Arthropods, 12, 51, 71, 72, 93, 100

Artifacts, Amerind, 19
Aruanã, 44, 45, 48, 113, 114
Atlantic Ocean, 4
Authorities, fish, 119
　Brazilian, 56
　local, 118

Babão, 28, 29, 72, 76–78
Bacu, 28, 87–89, 90
Bacu Comum, 28, 87, 89
Bacu Pedra, 28, 89–91, 92
Bacu Rebeca, 87–89, 90
Balance, nutrient, 118
Bandeirante, 17
Bank(s),
　left, 1, 4, 23, 29, 43, 78
　nutrient, 5
　right, 4, 38, 39, 56, 62
　terra firme, 4, 43
Barba-Chata, 28, 84, 88
Basin,
　Amazon, 1, 4, 17, 20, 23, 32, 52, 59, 74, 76, 78, 97, 98, 109, 117, 118
　　highway network of, 20
　　Southwestern, 23
　La Plata, 1
　Rio Madeira, see 'Rio Madeira'
Bates, 19
Beef, 24, 67
　imported, 24, 56
Behavior,
　feeding, 51, 69, 76, 78, 91, 98, 106
　　of most of the important food fishes of the Rio Madeira, 72
　　of *Mylossoma albiscopus* and *M. duriventris*, 98
Belém, 17, 23, 34
Bico de Pato, 28, 78, 79
Biomass, 62, 63, 71, 73, 84
　characin, 38

fish, 70
  commercial, 61, 62
  food fish, characin, 62
  high, 106
  of fishes, 54
  total, 62
Biotope, floating meadow, 71
Birds, water, 118
Blackwater(s), 5, 10, 43, 120
Bloodworms, 91
Boats, ice, 25, 44, 57
Bodó, 91, 93
Bogotá, Colômbia, 60
Bolivia, 22, 24, 56, 57, 58, 59, 72, 98
  eastern, 4, 59, 73, 74, 78, 120
Bony-tongues, 73
Boom, rubber, 22, 23
Bow(s)-and-Arrow(s), 25, 45–48, 67, 113
Branquinha cabeça lisa, 39, 105, 107
Branquinha chora, 39, 45, 105, 107
Branwuinha comum, 39, 45, 107
Branquinhas, 39
Brazil, 19, 22, 26, 57, 59, 60, 76
  Central, 23
  Northeastern, 23, 109
  Southeastern, 23, 26, 59, 60, 76
  Southern, 23, 26, 56, 59, 60, 76
Brazilians, 22
Brazilian Shield(s), 1, 2, 3, 4, 5, 8, 10, 12, 19, 23
Britski, Heraldo, 84

Caatinga, 10
Caboclo, 44, 54, 113
Cachoeira da Esperanza, 56
Cachoeira do Teotônio, 4, 6, 9, 26, 71
Cachoeiras, 26, 36
Cage, underwater, 57
Caimans, 118
Calama, 24, 76
Calhapo, 57, 58
Camurim, 48
Canal, escavation of, 19
Cará-açu, 44, 45, 109, 110
Cara de Gato, 83
Candirú-açu, 31
Candirú Pintado, 31
Caparari, 28, 81
Caribbean, 22
Caripunas, 23
Cascudinha, 39, 105, 107
Cassiterite, 16
Castnet, 25, 32–36, 45, 58, 76, 91, 113
Cataract(s), 3, 4, 5, 12, 16, 17, 18, 22, 26, 27, 28, 56, 67, 71, 109
  second major, 26
  largest of, 4
  Santo Antonio, 113

Teotônio, 27, 28, 29, 30, 31, 33, 34, 36, 67, 71, 74, 76, 78, 81, 84, 87, 89, 91, 103, 113
Catch(es),
  annual, 59, 60, 63
    history of, 60–63
    largest, 59
    monthly distribution of, 95, 104
    total, 60, 61, 63, 95, 118
  capacity, 56
  catfish, 60, 73
  commercial, 39, 43, 59, 72, 73, 83, 88, 97, 118, 119, 120
    most important, 28
  distribution of, 63, 64, 95
  freshwater, South American, 78
  low, 103
  Manaus, 97
  monthly, 66, 68
    distribution of, 63
    total, 66
  non-catfish gillnet, 52
  per unit of effort (CPUE), 63–67, 68, 118
  Porto Velho, 60
  potential, total, 118
  Rio Madeira, 67, 74
  seasonality in, 63
  siluroid, 60
  three year, 59, 65
  total, 63, 65, 67, 69, 72, 73, 113, 118
Catfish, 5, 25, 26–38, 48, 49, 52, 56, 60, 61, 67, 71, 72, 73, 97, 113, 120
  channel, river, 72
  commercial, 28
    most important, 28
  exploitation, 28
    firewood, 84
    large scale, 82
  important, most, 82
  migrating, upstream, 26, 27, 28, 29, 32, 34, 67, 71
  migration of, annual, 26
  neotropical, 78
  pimelodid, 80
  salted, 27
  schools of, 29
  tiger-striped, 81
  voracious, 29–32
  whale, 29
Cattle, beef, 24
Ceará, 22, 23, 24
Chains, food, 5, 12, 43, 51, 71, 72, 120
  alternative, 120
Channel(s), 4, 32, 38, 43, 92
  flowing, 39
  main, 8, 38, 81
  river, 3, 5, 6, 8, 36–38, 73, 76, 81, 93, 103, 118
Characin(s), 26, 27, 32, 36, 38, 42, 43, 49, 52, 54, 56, 62, 70, 72, 73, 113, 120

baited with, 38
deep-keeled, 100
food-fish, 38, 92, 119
fruit eating, 49
migrating, 69
   upstream, 4, 36
migratory, 38, 39, 42, 43, 54, 60, 63, 67, 68, 70, 71, 72, 93, 97, 103, 105, 119, 120
   adult, 62
   spawning, 39
non-predatory, 62
schools of, 42, 83
spawning, 39, 54
Chicken, for catfish exchange, 25
Chico Duro, 105, 107
Church, George, 22
Cichlid(s), 26, 44, 48, 52, 73, 109
   African, 118
   deep-bodied, 109
   exotic, 118
Clearwater(s), 10, 12, 16, 41, 43, 56, 120
Coroatá, 83
Clupeid, predatory, 113
Community(-ies), 69, 117, 119
   fish, 117
macrophyte, 71
   plankton, 72
Collector(s),
   data, 63, 98
   latex, 23
   rubber, 23, 27
Company(-ies), refrigeration, 24, 25, 26, 27, 28, 56, 60, 61, 63, 80, 91
   Brazilian, 26, 56, 58, 59, 98
   Porto Velho, 76
Concentrations, suspended solids, 10
Contents, stomach, 88
Control, types of fishing gear, 119
Cooperation, international, 120
Cooperative, Porto Velho fishermens', 63
Cordillera, Andean, 5
Coronel, 83
Covo, 32, 33
Crab, 113
Croakers, 113
Crop, agriculture, 16
   modification of floodplains for, 120
Crown, Portuguese, 17
Cuiu-Cuiu, 28, 87, 89
Culturalists, Amazonian fish, 118
Culture,
   experimental, 87
   fish, 118
Cuniã, 9, 44, 48, 49
   fisheries, 44
   fishermen, 44, 45, 48, 50

floodplain, 46, 47
Lago de, 43, 50, 53
Curimatá, 39, 45, 54, 56, 57, 100-104
Curimatid(s), 54, 103, 106
   detritivorous, 44
Currico, 48
Curumim, 49
Curumim-line, 45, 49, 50
Cucurbits, 51
Cycles, life, 120

Dam(s), 16, 117
DC-3 (aircraft), 24, 57
Deforestation, 16
   large-scale, 117
   of the Rio Madeira floodplain, 16
Deforesting, 120
Debouchures, affluent, 25
Deepwater-gillnet, drifting, 36, 37, 38, 73, 76, 78, 81, 82, 89, 91
Deposits,
   alluvial, 8
   fat, 42
   lacustine, 4
Detritivores, 63, 70
Detritus, 48, 71, 72, 78, 88, 91, 106
Dipnets, 42
Discharge,
   annual (Rio Negro), 1
   of Amazonian rivers, 3
   total, 4
      Madeira's, 1
Dispersal, seed, 12
Dolphins, 42, 118
Doradid(s), 87, 89
Dourada, 27, 28, 29, 30, 38, 60, 72, 72–75, 76, 92
Dourada Fita, 28, 84, 86
Dourada Zebra, 84, 85
Drainage,
   Amazonian (system), 5
   Bolivian, eastern, 4
Draining, lake, 50
Ducke, Adolpho, 19

Eaters, seed and fruit, 98
Ecology, Amazonian fisheries, 12, 119
Ecosystem(s), 16, 69
   aquatic, 16, 51
   natural, 5, 16
   overall role of predators in, 118
Eel, electric, 73
Effort,
   catch per unit of, 63
      in terms of fuel, 67
      in terms of ice, 68
      in terms of man-days, 65–67

commercial, 98
fishing, 60, 61, 63, 69, 106, 109
   commercial, 39
   total, 63
  subsistence, 98
  total, 65, 67
Empires, Portuguese and Spanish, 17
Erosion, 5, 8, 12, 16
Espírito Santo, 23
Estuary, Amazon, 76
Europe, 22
Expedition(s), 17, 19
  Roosevelt-Rondon, 19, 74
  scientific, 17
  Tavares, 17
Exotic species, introduction of, 118
Exploitation,
  commercial, 51, 83, 98
  heavy, first years of, 81
  large-scale, 82
  of fishes, 70, 119
  of *jatuarana*, 92, 95
  of spawning *jatuarana*, 95
  over-, 63
  siluroid, large-scale, 84
  vulnerable to, 27
Exploration (of the Rio Madeira Basin), 17–22
  scientific, 22
Explorer, 19
Extinction, threatened with (*Piracucu*), 113

Factor, main environmental, 69
Farinha, 44
Farmers, subsistence, 46, 67
Farquar, Percival, 22
Fat, body, 72
Fat deposits, 42
Fat reserves, 70
Fat stores, 70
Fauna(s)
  Amazon, 69
  fish, 1, 63
    freshwater, 59
      world's most diverse, 69
    natural, 118
  piscine, 17
Feeders, microphagous, 39
Ferry, 21, 23
Ferreira, Alexandre Rodrigues, 17
Fields, cassiterite, 27
Filhote, 28, 45, 49, 74–76
Fisga, 29
Fish(es),
  commercial, 48, 72, 73
    food chain sustaining the, 72
  detritus eating, 60, 72

detritus feeding, 72, 103
electric, 73
exportation of, 25
floodplain, 109, 113
food, 26, 45, 57, 60, 72–73, 84, 88, 92, 100, 105, 106, 109, 113, 118
  commercial, 73, 98
  first class, 43, 92, 109, 119, 120
  important, 60, 72, 73, 91–92
  one of the best, 113
  Rio Madeira, 60
  second class, 103
  third class, 113
frugivorous, 51
fruit eating, 51
fruit and seed eating, 100
marketable, 63
microphagous feeding, 42, 103
migratory, 92
piscivorous, 73
prey, 62
red, 109
scaled, 56
  second largest, 98
  largest, 97
second class, 76, 103, 106
tropical, 118
upstream migrating, 4
Fisherman(-men),
  Bolivian, 56, 57, 58, 98, 103
  Brazilian, 57, 103
  commercial, 4, 38, 39, 76, 119
  Cuniã, 44, 46, 48, 50
  gillnet, drifting deepwater-, 89
  Guajará-Mirim, 57
  Itacoatiara, 65
  pole, 51
  Porto Velho, 36, 44, 61
  Rio Guaporé, 57
  Rio Madeira, 18, 26, 47, 60, 63, 81
  Rio Mamoré, 57
  seine, 42
  skilled, 25
  subsistence, 83, 84, 119
  Teotônio, 27, 29
  weir, 54
Fishermen-miners, 27
Fisheries, The, 26, 57
Fishery(-ies), 12, 24, 25, 26, 27, 36, 38, 41, 44, 51, 54, 56, 63, 65, 67, 72, 76, 98, 118
  Amazon(-ian), 12, 67, 118
  annual, 43
  cataract, 67, 78, 88
    Teotônio, 61, 66
  catfish, 26–29
  characin, migratory, 38–43

commercial, 26, 38, 39, 42, 44, 56, 65, 67, 68, 72, 73, 89, 91, 97, 100, 103, 106, 109, 119
    Amazonian, 25
Cuniã, 44, 66, 67, 109
direct management of, 63
evaluation of, 63
feeding, microphagous, 42
flooded forest, 51, 54, 67
    nature of, 25
floodplain, 42, 43–50, 67, 119
frugivorous, 51
gaff, 29
Itacoatiara, 65
natural history of, 117
nature of, in relation to urban centers, 25
pirarucu, 44
predatory, 48
prey, 62
Rio Amazonas, 65
Rio Madeira, 12, 18, 25, 26, 38, 41, 44, 47, 54, 67, 91, 92, 98, 106, 109, 118, 119
    relative productivity of, 117–118
river, 117
    African, 63
    management of, 117
seine, 66, 67, 113
siluroid, 26
subsistence, 25, 73, 87, 97, 109
weir, 54, 56
Fishing
    commercial, 27, 51, 61, 62, 71, 97
        large scale, 59
    low water, 56
    pole, 49, 51
    protection against large-scale, 54
    second most productive period of, 63
    seine, 54
    subsistence, 32
    techniques of Rio Madeira floodplain, 44
Fishing effort, see 'effort'
Fishing gear, control types of, 119
Fishmongers, 24, 25, 57, 81, 92, 98
Flecheiro, 39
Fleet (fishing),
    Itacoatiara, 59
    Porto Velho, 59, 60, 64, 66, 95, 103, 118
Float(s), 37, 42, 45, 49
'Floating meadows', 12, 71, 81
Flood(s), 1, 4, 7, 10, 11, 12, 27, 28, 38, 39, 42, 44, 46, 51, 56, 88, 93, 97, 98
    annual, 1, 10, 38, 44, 92, 106, 119
    beginning of, 106
    height of, 63, 70
Flooded forest(s), 4, 5, 8, 10, 12, 13, 25, 41, 42, 43, 44, 45, 49, 51, 52, 54, 62, 63, 65, 70, 71, 72, 73, 78, 81, 88, 91, 92, 98, 100, 103, 106, 113, 117, 119, 120

clearwater, 12
Floodplain(s), 4, 5, 7, 8, 9, 10, 12, 16, 24, 43, 44, 45, 50, 54, 62, 63, 70, 72, 88, 91, 98, 109
    Amazonian, 14
    blackwater, 12
    Cuniã, 43, 44, 46
        fisheries, 67
    river, turbid, 12, 120
    Solimões, 98, 106
Floodplain lakes, see 'Lakes'
Floodplain systems, see 'Systems'
Flow,
    protein, animal, 24, 25
    seasonal, of the Rio Madeira, 12
Fluctuation, river level, 10–12
Folklore, local, 80, 109
Fonseca, Gonçalves da, 26
Food chains, see 'Chains'
Fork length, 31, 73, 74, 76, 77, 78, 79, 82, 83, 84, 87, 88, 89, 91, 94
Frenchmen, 22
Frontier, new, 56–58
Fruit(s), 51, 70, 71, 72, 78, 88, 89, 91, 93, 97, 98, 100, 106
    palm, 51, 97
    rubber tree, 97
Fry, 71
Fuel, catch per unit of effort in terms of, 67

Gaff(s), 29–32, 78
Gaff-hook, 29
Gaponga, 51
Gas, hydrogen sulfide, 48
Gear,
    control types of fishing, 119
    dominant, 25
    traditional, 67
Genera, food fish, 91
Geology, surface, 5
Gig(s), 25, 44, 45, 56, 67, 109
Gillnet(s), 25, 44, 45, 52, 54, 59, 97, 98, 100, 109, 113, 119
    cotton, 54
    drifting beach, 55, 56
    drifting deepwater-, 36, 37, 38, 73, 76, 78, 81, 82, 89, 91
    flooded forest, 52, 53
    large-meshed, 119
    mesh, 97
    monofilament, 54
    non-catfish, 52
    nylon, 54
    small-meshed, 119
Gnerium, 47
Gold, 17
    wiskered, 27
Government,
    Bolivian, 19, 117
    Brazilian, 19, 22, 26, 57, 117
    federal, 27

local, 25
of Amazonas, 25
Grounds, spawning, 78
Groups, Amerind, 23, 54
Grozeira, 36, 38
Guajará-Mirim, 4, 23, 24, 56, 57, 58, 59, 98
Guayaquil, Gulf of, 1
Gudger, 74
Guiana Shield, 1, 5
Gulf of Guayaquil, 1
Gymnotoids, 73

Habitats, nursery, 71, 76, 120
Handline(s), 32, 33, 45, 48, 113
Harpoon, 25, 45, 49, 51, 109
Headwaters, 1, 5, 19, 23
High water level, see 'Level'
High water season, see 'Season'
Highway(s), 23
   Amazonian, 24
   Porto Velho/Cuiabá (Cuiabá/Porto Velho), 16, 23, 24, 26
   Porto Velho/Manaus, 16, 21, 23
   Transamazonian, 16, 23, 24
History,
   geographical, 1–5
   natural, 19
   of annual catches, 60–63
   transportation, Amazonian, 22
Hoehne, F.C., 19
Humaitá, 19, 23, 25
Hydrochemistry, 1, 5–10
   of the Rio Madeira, 5

Ice, 25, 56, 57, 68, 106
   catch per unit of effort in terms of, 68
Ichthyofauna, commercial, 72
Ichthyology, Amazon, 19
Igapó, 12, 54
Ihering, H. von, 19
Image, Landsat satellite, 9, 19
Indians, 17, 27
Insecticides, 120
Insects, 78, 109, 113
Inundation,
   annual, 27, 88, 91, 103, 119
   normal, 62
Itacoatiara, 24, 36, 59, 65, 67, 74

Jaraqui escama grossa, 39, 45, 103–105
Jaraqui escama fina, 39, 103–105
Jatuarana, 91–93, 94, 95, 103
Jaú, 28, 32, 33, 35, 38, 78–81
Jauari, 51, 103–105
Jiparaná, 16, 24

Keller, Franz, 18, 19, 26, 27

Lago de Cuniã, 43, 50, 53
Lagoons, floodplain, 44, 54, 62, 71
Lake(s),
   Bolivian, eastern, 4
   floodplain, 10, 25, 38, 44, 54, 62, 71, 73, 88, 103, 109, 113, 119
Lake draining, 50
La Paz, 1
La Plata, 78
Larvae, insect, 88, 91
Leaves, 51, 98, 106
Levees, alluvial, 4, 54
Level, river, 39
Level(s),
   water, 4, 42, 54, 60, 64, 91, 95, 97, 104
   differential, 4
   ecosystem, 5
   high, 12, 95, 104
      maximum, 62, 63
      monthly, 64
   low, 12, 28, 62, 95, 104
      extremely, 60
      minimum, 12
   minimum, 46
   nutrient, 10, 12
   rising, 54
   river, 39
Limnology, 71
Line-and-pole, 46
Line of Tordesillas, 17
Line, telegraph, 19
Littoral, Atlantic, 17
Load(s), suspended, 5, 8, 16
Loricariids, 91
Lowlands,
   Amazon, 1, 5, 8, 12, 19
   rainforest covered, 5
Low water level, see 'Level, water'
Low water period, see 'Period'

Macrophyte(s), 10, 43, 44, 51, 71, 72, 91
   aquatic, 10, 12, 72, 88
Madeira-Mamoré Railway, 3, 21, 22, 23
Malaria, 22, 24
Manacapurú, 25
Manaus, 24, 36, 52, 59
Man-day(s), 63, 65
Mandi, 28, 78, 79
Manicoré, 24
Manioc, 44
Mapará, 91, 94
Market,
   Manaus, 59, 60, 64, 65, 67, 97
   Porto Velho, 54, 60, 62, 78, 98, 100, 106
Martins, 19
Mato Grosso, 17, 23, 24, 27
Matrinchaõ, 93, 96

Mayflies, 88
Methods, indigenous fishing, 25
Microfaunas, 43
Midges, 88
Migration(s), 27, 28, 38, 39, 43, 69, 70, 92, 97, 103, 120
    annual, first, 103
        of *Prochilodus nigricans*, 103
    catfish, 27, 71
        upstream, 27, 71, 97
    characin, 69
    dispersal, 62, 119
    fish, 66, 69
    hypothetical, 119
    large-scale, 103
    main, 92
    Semaprochilodus, 106
    spawning, 38, 39, 41, 92
    upstream, 27, 42–43, 62, 72, 76, 103
Migration patterns, see 'Patterns'
Mining, tinstone (cassiterite), 9, 15, 16, 27
Miocene, 1
Mississippi, 1
Modification, human (of the Rio Madeira Basin), 16
Months, low water, 27
Morphology, river, 1–5
Museu Nacional (Rio de Janeiro), 19
Mutum-paraná, 3

Names, Amerind, 17
Negresses, 22
Negroes, 22
New frontier, 56
Novo Aripuanã, 24
Nutrient(s), 5, 43, 71, 119
    poor in, 43
Nutrition, fish, 120

Ocean,
    Atlantic, 4
    Pacific, 1
Officials, government,
    Bolivian, 16
    Brazilian, 16
Operations, logging, 16
    large-scale commercial, 92
Orana, 39, 106, 110
Orogeny, Andean, 1
Over-exploitation, 64, 73
Overfishing, 44

Pacific (Ocean), 1
Pacu branco, 39, 43, 98, 99
Pacu encarnado, 39, 45
Pacu mafurá, 98, 101
Pacu toba, 39, 45, 98
Pacu vermelho, 39, 98, 99

Pantanal, 1, 24
Paraná, 23, 25, 26
Patterns, migration, 95, 106, 113
Pauzadas, 97
Peak,
    flood, 4, 12
        of the Rio Madeira, 11
    flood level, 10
Peixe Gordo, 42, 70
Peixe Lenha, 28, 84, 86
Period (season),
    flooding, 4, 5, 11
        pre-spawning, 38
    low water, 4, 7, 11, 12, 17, 27, 28, 32, 36, 44, 46, 48, 50, 56, 57, 62, 63, 66, 68, 69, 71, 76, 78, 81, 82, 84, 88, 89, 91, 93, 97, 98, 100, 103, 106, 109, 113, 120
    Quaternary, 1, 4
    spawning, 39, 98, 106, 119
    Tertiary, 1
Periphyton, 43
Perizoon, 43
Pescada, 113, 115
Petropolis, Treaty of, 22
Phytoplankton, 5, 10, 43, 51, 118
Pimelodids, 80
Pindá-lure, 46
Pintadinho, 28, 87, 88
Pião, 39, 106, 108
Pindauaca, 46
Piracatinga, 87, 88, 98
Piracema, 43, 61, 63, 68, 70, 93
Piraiba, 28, 32, 38, 40, 74–76
Piramutaba, 28, 38, 76, 77
Piranha(s), 19, 44, 52, 56, 97, 100, 119
Piranha Caju, 45, 49, 100, 102
Pirapitinga, 39, 45, 49, 97, 99
Pirarara, 28, 38, 83–84, 85
Pirarucu, 44, 45, 49, 74, 109–113
'Piscine miners', 27
Plains, eastern Bolivian, 12
Plankton, 43, 44, 71
Plantations, East Asian, 22
Plant, thermoelectric, 16
Pliocene, 1
Poisoning, vitamin A, 76
Poisons, fish, 27
Pole-and-line, 25, 45, 51, 100
Pollution,
    ecological effects of, 16
    industrial or agricultural, 5
    stream and river, 16
Population (of the Rio Madeira basin), 23–24
Porto Velho, 3, 4, 10, 11, 22, 23, 24, 25, 26, 27, 36, 43, 44, 54, 56, 57, 59, 67, 76, 81, 88, 91, 98, 105, 109
    foundation of, 22
Portuguese, 17, 22

Power, hydroelectric, 16
Predation, 61, 62
  vulnerable to, 63
Predator(s), 29, 31, 62, 109, 113, 118
  abundant, one of the most, 100
  fish, extermination of, 118
  piscivorous, 118
  sharp-toothed, 56
  smallest, 29
Production,
  fish, 118
    surplus, 117
  plankton, 71, 91
    very poor, 43
Prospecting, tin-stone, 27
Protection, environmental, 120
Protein, animal, 24, 25, 65, 91, 117
P.T. Collins (American engineering firm), 22

Quaternary Period, 1, 4

Railroad, see 'Railway'
Railway, 19
  Madeira-Mamoré, 3, 21, 22, 23
  rainforest, construction of, 22
Rainfall, 8, 10
Rainforest, 10, 12, 21
  tropical, 12
Raiva, 46
Ranching, cattle, 16
Randam, 19
Raposo, Antonio Tavares, 17
Rapids, 4, 28, 32, 36, 59, 67, 71, 76
  Madeira, 4, 11, 19, 22, 26, 27, 29, 76, 78, 80, 87, 103, 113
  Santo Antonio, 27
  Teotônio, 26, 27, 28, 31, 32
Recycling, nutrient, 118
Rede de lance, 41, 42
Refrigeration companies, see 'Company'
Region(s)
  fisheries, Central Amazon, 59
  fisheries of the Rio Madeira, 63
  fishing, three main commercial, 59
  natural history of, 19
  savanna, 4, 5
Relatórios, 17
Remoso, 76
Reports, 17
Reserves, fat, 70
Resource, untapped fish, 118
Restrictions, gear, 119
Ribeiro, Alípio Miranda, 19
Rio Abunã, 12
Rio Amazonas, 1, 4, 9, 10, 11, 12, 17, 24, 36, 59, 65, 67, 70, 73, 74, 76, 120
  annual volume of, 1

Rio Aripuanã, 4, 10, 16, 36, 43, 59, 63, 70, 73
Rio Beni, 4, 11, 12, 24, 56, 59, 73, 78, 103, 117
Rio Branco (Acre), 24, 54, 58, 59, 73, 78, 103, 117
Rio Canumã, 10
Rio da Dúvida, 19
Rio Grande do Sul, 23
Rio Guaporé, 1, 17, 24, 26, 56, 57, 58, 59, 82, 98, 103, 109, 117
Rio Jamari, 10, 16, 56, 70, 91, 113
Rio Jancundá, 16
Rio Jiparaná (Machado), 19
Rio Juruá, 22
Rio Machado, 8, 10, 14, 16, 19, 41, 56, 70, 78, 113
Rio Madeira,
  affluents, 4, 5, 19, 23, 38
  backdrop of, cultural, 17
  basin, 15, 16, 17, 19, 22, 23, 24, 25, 26, 36, 38, 39, 62, 69, 70, 98, 100, 109, 113
    drainage, 12
    fish migrations in the, 69–72
    human modification of the, 16
    population of, 23–24
  cataracts, 3, 22, 26
  channel(s), 27, 32, 37, 39, 40, 43, 55
  cultural backdrop of, 17
  data,
    catch, 62
    river level, 12
  data for, water level, 10
  discharge, total, 1, 4
  drainage system, 2, 10, 12, 59
  fisheries, see 'Fishery'
  fisherman(-men), see 'Fisherman'
  fishes, food, 60
  flood(s), 11, 12
  flooded forests of, 12, 13, 51
  floodplain(s), see 'Floodplains'
  flow of, seasonal, 12
  fluctuation of, 10, 11
    river level, 10
  high water level of, 12
  high water season of, 10
  history of, natural, 19
  hydrochemistry of, 5
  level of, high water, 12
  lower, 9, 10, 56, 59, 74, 120
    Landsat satellite image of, 9
  low water period(s), 7, 12
  middle, 60, 70, 71, 72, 103
  migratory characins of, 39
  modification of, human, 16
  natural history of, 19
  periods, low water, 7, 12
  rapids, see 'Rapids'
  region, 26, 46, 52, 62, 63, 72, 73, 80, 91, 98, 103, 105, 109, 119
    floodplain, 70
  regions of, fisheries, 59

river level fluctuation of, 11
river level data, 12
seasonal flow of, 12
season of, high water, 10
sediment of, 4
situation of, geographical, 1
system, drainage, 2, 10, 12, 59
tributaries of, 10, 56
turbid, 38, 63
turbidity of, 8, 38, 39, 41
turbid water(s) of, 36, 38, 39, 54, 98
upper, 3, 6, 9, 11, 16, 54, 55, 56, 60, 70, 71, 72, 103
    Landsat satellite image of, 9
valley, 23, 24, 26, 27, 51, 56, 91, 97, 98, 117
    animal protein flow in the upper, 24–25
    flooded forests of, 51
volume, annual, 1
water level, 12
    annual fluctuation in, 63
water level data for, 10
watershed of, 12
water(s) of, turbid, 36, 38, 39, 54, 98
Rio Mamoré, 4, 11, 12, 17, 22, 26, 56, 57, 58, 59, 73, 78, 82, 98, 103, 117
Rio Manicoré, 10
Rio Marmelos, 10
Rio Negro, 1, 10, 19, 59, 105
    discharge, annual, 1
Rio Pilcomayo, 24, 57
Rio Preto do Igapó-Açu, 10, 16
Rio Purus, 1, 4, 22, 23, 43
Rio Solimões, 4, 23, 71, 76
Rio Solimões-Amazonas, 4, 5, 10, 19, 24, 38, 43, 44, 71
    total drainage area, 1
Rio Tapajós, 10, 19
River(s),
    Amazon, 17, 112, 113
        comparative discharges of, 3
    Andean, 4
    blackwater, 10
    Bolivian, 72
        eastern, 3, 4
    clearwater, 8, 16, 51, 56, 113
    floodplain, 117
    Madeira, 112
    nutrient poor, 118
    polluted, 16
    São Paulo, 17
    tropical, 118
    turbid, 5, 12, 43
    turbid water, 43, 71
Rock, Ceará, 34
Rodada, 39
Rondon, Candido Mariano da Silva, 19
Rondônia, 15, 16, 23, 24, 25, 59, 118
Roosevelt, Teddy, 19, 74

Roots, macrophyte, 106
Rubber boom, 22, 23
Runs, spawning, 92, 119

Salinity, 5, 10
Salto de Theotonio, 27
Salts, 5
Samoel Cataract, 16
Santo Antonio rapids, 27
São Paulo, 17, 25, 26, 27, 78
Sardinha, 100
Sardinha chata, 39, 100, 101
Sardinha comprida, 39, 100, 101
Satellite, Landsat, 19
Savannas,
    Bolivian, 5, 12
    swampy, 5
Scales, red, 109
Schools (fish), 38, 42, 54, 56, 63, 68, 69, 70, 76, 78, 83, 84, 88, 89, 91, 97, 100, 106, 113
    characin, 70, 83
    migrating, upstream, 78, 98
    of *jaraqui*, 103, 105
    of *pescada*, 113
    tambaqui, 97
Scientific expedition, 17
Season,
    closed, 119
        built-in, 119
    dry, 8, 23
    flooding, 4, 5, 11
    water, high, 12, 25, 28, 63, 72, 73, 100
    of the Rio Madeira, 10
Secchi Disk Transparency (SDT), 5, 10
Sediment(s), 4
    lacustrine, 3
Seeds, 51, 54, 70, 71, 72, 89, 91, 93, 97, 98, 100
    tree, rubber, 51, 52
Seine(s), 25, 41, 42, 43, 54, 56, 57, 62, 63, 67, 73, 78, 83, 87, 88, 97, 100, 113, 119
    beach, 43
    open water, 43
    small-meshed, 119
Seringal, seringais, 23
Serra Tres Irmãos, 3, 4
Shield(s),
    Brazilian, 1, 2, 3, 4, 5, 8, 10, 12, 19, 23
        western, 1, 23
    Guiana, 1, 5
Shrimps, 88, 113
Siluroid, 26, 29, 62, 71, 72, 73, 74, 84, 120
Sites,
    dam, 16
    tree, rubber, 23
Slopes,
    Andean, western, 5

Cordilleran, 5
Soils, acidity of the granitic and gneissic, 10
Sorubim, 82
Source, animal protein, 24
    alternative, 24, 25, 117
South America, 17, 19, 74, 78
Spaniards, 22
Spawning, 39, 72, 78, 92, 97, 98, 106, 119, 193
Species, introduction of exotic, 118
Spiders, 113
Spix, 19
Spruce, 19
Standard length, 81, 88, 91, 93, 94, 96, 97, 98, 99, 100, 101, 102, 103, 104, 106, 107, 109, 110, 111, 115
Stomach content analyses, 72, 73
Stores, fat, 70
Streams, rainforest, 25, 54
Supply, protein, 91
    animal, 65
Surubim, 27, 28, 82
Survey(s), gillnet and seine, 73
System(s),
    Amazon, 19
    drainage, 1, 59
        Amazonian, 5
        Rio Madeira, 2, 10, 12
            fisheries regions of the, 59
    floodplain, 44
    natural, 118
    Paraguay-La Plata, 1
    Rio Madeira as a, 17
    river, 19, 69
        blackwater, 43
        great, 19
        nutrient poor, 43, 118, 120
        tropical, 43
        turbid, 43
    transportation, 19
    tributary, 103
    water, nutrient enriched, 5

Tambaqui, 39, 42, 43, 45, 49, 51 56, 96, 97–98
Tapagem, 54
Tarrafa, 48
Taxa, 42, 70, 78, 92
    characin, 69
    commercial, 119
    common, 100
    fish, 56, 72, 92, 98
    frugivorous, 69, 70
    top ten, 60
Taxonomy, 78, 91, 100, 106
Teotônio, 36
Terra firme, 4, 8, 10, 41, 43
Territory, Acre, 22
Tertiary Period, 1

Tilapia, 118
Through the Brazilian Wilderness (Roosevelt 1914), 19
Tordesillas, Line of, 17
Traíra, 44, 45, 106, 110
Trailblazer, Portuguese, 17
Transparancy,
    good 62
    Secchi Disk (SDT), 5, 10
    water, 5, 16, 46
Trap(s),
    catfish, Teotônio, 32
    insecticide, 120
    nutrient, 5
    wire (*covo*), 32, 33
Treaty of Petrópolis, 22
Tree(s),
    cotton, kapok, 12, 14, 16
    rubber, 22, 23, 51, 52
Tributary(-ies),
    Andean, 5, 8
    blackwater, 10, 119, 120
    clearwater, 38, 55, 56, 62, 70, 71, 83, 87, 88, 92, 93
    poor, nutrient, 71, 120
    rightbank, 4, 10, 12, 16, 23, 38, 39, 63, 70, 72, 109
    water, turbid, 19
Trotline, 37, 38, 40, 45, 52, 76, 84, 87
    baited, fruit and seed, 52
    channel, river, 36–38
    flooded forest, 52
Tucunaré, 44, 45, 46, 48, 50, 56, 72, 100, 109, 110
Turbidity,
    of the clearwater rivers, 16
    of the Rio Madeira, 8
Turtle, river, 57

United States, 22
Urubus (black vultures), 27

Várzea, 44
Vegetation, 12, 19
Vilhena, 16, 24
Vultures, black, 27

Wallace, 19
Water(s),
    Amazonian, 78
    Argentinian, 78
    Bolivian, 120
    Brazilian, 120
    enriched, nutrient, 5
    high, 10
    low, 10, 93
    transparancy, low, 39
    turbid, 39, 41, 43, 54, 98
Water level, see 'Level'
Water transparancy, see 'Transparancy'

Watershed(s), 12, 16
Weir(s), 54, 55, 56
   fish, 55

Year, low water, 62
Yield(s),
   by species, 59–60
   in comparison to floodplain area, 117–118
   man-day, 65, 67
      mean, 65, 67

      annual, 67
   Rio Amazonas, 67
   Rio Madeira, 67, 118
   sustained, 63
      maximum, 117

Zagáia, 44
Zaire, 1
Zooplankton, 63, 97, 100, 118
   crustacean, 91